N-dimensional crystallography

R L E Schwarzenberger
University of Warwick

N-dimensional crystallography

Pitman Advanced Publishing Program
SAN FRANCISCO · LONDON · MELBOURNE

PITMAN PUBLISHING LIMITED
39 Parker Street, London WC2B 5PB

North American Editorial Office
1020 Plain Street, Marshfield, Massachusetts 02050

North American Sales Office
FEARON PITMAN PUBLISHERS INC.
6 Davis Drive, Belmont, California 94002

Associated Companies
Copp Clark Pitman, Toronto
Pitman Publishing New Zealand Ltd, Wellington
Pitman Publishing Pty Ltd, Melbourne

AMS Subject Classifications: (main) 20H15
(subsidiary) 50B30, 82A60

Library of Congress Cataloging in Publication Data

Schwarzenberger, R L E
N-dimensional crystallography.

(Research notes in mathematics)
Bibliography: p.
1. Crystallography, Mathematical. I. Title.
II. Series.
QD911.S35 548'.7 79-19693
ISBN 0-8224-8468-4

Manufactured in Great Britain.

US ISBN 0-8224-8468-4
UK ISBN 0 273 08468 2

Preface

There are three main justifications for the study of symmetry groups in n-dimensional euclidean space.

First, as a particular chapter in the theory of discrete subgroups of a Lie group (in this case the Galilean group of all isometries of n-dimensional space). Chapters Two and Three of these notes form an introduction to this point of view.

Second, as a contribution to the theory of finite groups. If it were possible to obtain a complete enumeration of n-dimensional symmetry groups for all n = 1,2,3,.... then in particular there would be complete information on all finite groups and on their integral representations. Chapter Four and Five contain some partial results in this direction.

Third, as an approach which clarifies understanding of the situation in particular low dimensions, especially n = 2,3,4. A self contained description of results for n = 2 (the 17 *wall paper* and 46 *black and white* groups) is given in Chapter One. More general results, which illustrate how the number of n-dimensional symmetry groups grows as n increases, can be found in Chapter Six together with an explicit enumeration of the 230 groups for n = 3 familiar to crystallographers and of particular examples for n = 4.

The approach adopted is geometric, and the choice of topics is intended to complement those accessible via the alternative algebraic approach. The latter is associated with Zassenhaus and his collaborators and yields a computational algorithm for the determination of all n-dimensional space groups (a programme carried out successfully for n = 4, when there are 4783 distinct groups, but unlikely to be implemented for larger n because of the huge number of groups involved). Further significant results have been obtained with Lie algebra techniques by Maxwell, Plesken and Pohst. However the geometric approach seems better adapted to an elementary treatment and to the search for qualitative results valid in arbitrary dimensions. It is in the hope that such results will be found in the near future that these notes have been written.

A large part of this material has been included in undergraduate courses given at the University of Warwick, and improved by comments from the audience. The final version is based on seminars given at the University of Nagoya under the Monbusho/British Council exchange scheme. For particular help and encouragement I am grateful to Akihiko Morimoto at Nagoya, to John Jarratt at Auckland and to Ian Stewart and Sue Tall at Warwick.

University of Warwick

August 1979 Rolph Schwarzenberger

Contents

1 Symmetry in the plane

These lectures are concerned with groups of symmetries in n-dimensional space. Some readers will, because of the applications to chemistry and physics, be concerned particularly with the case $n = 3$, and many examples will be worked out for this case in later chapters. All the essential concepts, however, occur already for $n = 2$ and can then be illustrated by patterns in the plane. It is therefore helpful to understand the concepts in this easy case, which is the subject of this introductory chapter. Readers who prefer a more algebraic treatment - without pictures - may turn straight to the next chapter where the more formal development of n-dimensional symmetry begins.

1.1. GROUPS OF ISOMETRIES

We consider symmetries (v,ϕ) which map

$$x \longmapsto v + \phi x$$

where v is a translation vector in the plane, and ϕ is a rotation (if $\det \phi = +1$) or reflection (if $\det \phi = -1$). Then

$$(v,\phi)(w,\psi)\, x = (v,\phi)(w+\psi x) = v+\phi(w+\psi x) = v+\phi w+\phi\psi x$$

and hence

$$(v,\phi)(w,\psi) = (v+\phi w,\ \phi\psi),$$

$$(v,\phi)^{-1} = (-\phi^{-1}v,\ \phi^{-1}).$$

Let G be a group consisting of such symmetries. The structure of G involves (and for certain types of group will turn out to be completely determined by) four invariants, as follows.

(i) The lattice T of G. This is the set of translation vectors t such that $(t,\iota) \in G$ where ι denotes identity transformation. Since $(s,\iota)(t,\iota) = (s+t,\iota)$ this is an abelian subgroup of G, and also of the vector space R^2.

(ii) The point group H of G. This is the set of all linear transformations ϕ such that $(v,\phi) \in G$ for some $v \in R^2$. The set of all rotations in H forms a subgroup $H_o \subseteq H$.

(iii) The action of H on T. If $\phi \in H$ and $t \in T$ then for some $v \in R^2$ the element

$$(v,\phi)(t,\iota)(v,\phi)^{-1} = (v+\phi t)(-\phi^{-1}v,\phi^{-1}) = (v+\phi t-v,\phi\phi^{-1}) = (\phi t,\iota)$$

is in G and therefore $\phi t \in T$.

(iv) The shift vectors of H. Let $\phi \in H$ be of finite order q. Then

$$(v,\phi)^q = (v+\phi v+\ldots+\phi^{q-1}v,\phi^q) = (v+\phi v+\ldots+\phi^{q-1}v,\iota)$$

and therefore the element ϕ determines a translation vector $a = v+\phi v+\ldots+\phi^{q-1}v$ called the shift vector of ϕ. It is not defined uniquely: if v' is another choice for v then

$$(v,\phi)(v',\phi)^{-1} = (v,\phi)(-\phi^{-1}v',\phi^{-1}) = (v-v',\iota)$$

and so $v-v' = t \in T$ and $a-a' = t+\phi t+\ldots+\phi^{q-1}t$. In fact if ϕ is a rotation then $a = 0$, while if ϕ is a reflection in some line ℓ then $a = v+\phi v$ is an element of T which lies on ℓ. The diagram illustrates (v,ϕ), for different possible values of a, as v is varied by multiples of $t \in T$.

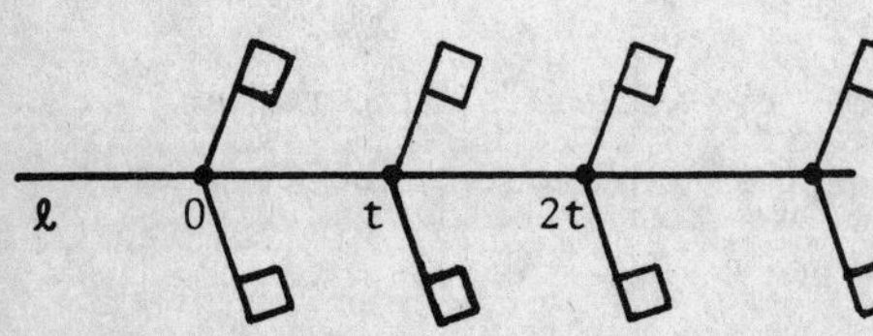

a = 0, 2t, ...

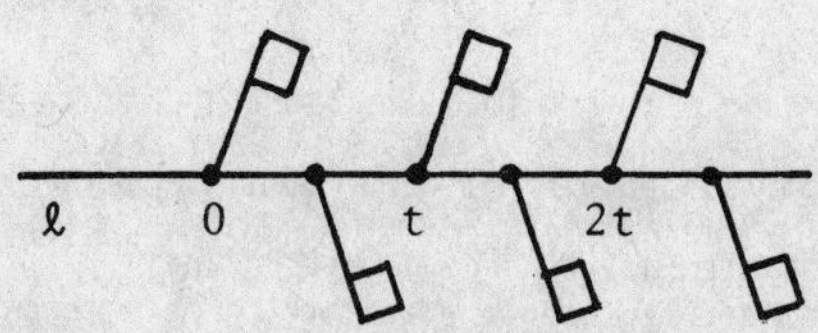

a = t, 3t, ...

In the next section attention is restricted to a particular type of group G, and the consequences of this restriction for the invariants (i)-(iv) are derived.

1.2. PLANE GROUPS

We shall obtain a classification theorem for groups G which satisfy

Definition G is a *plane group* if there are linearly independent vectors t_1, t_2 such that the lattice of G has the form

$$T = \{n_1t_1+n_2t_2;\ n_1n_2 \text{ integers}\}$$

and if moreover the point group H of G is finite.

The consequences of this definition for the invariants (i)-(iv) are

(i) T *contains a (not necessarily unique) non-zero vector* t *of minimum length* $|t|$. Consider for $c > 0$ the pairs of integers (n_1,n_2) such that $|n_1t_1+n_2t_2| < c$; being finite in number they yield a vector of minimum non-zero length.

(ii) H_o *is cyclic with generator a rotation* θ *through* $2\pi/q$ *for some integer* $q > 0$. Choose q the largest integer such that $\theta \in H_o$.

(iii) $q = 1,2,3,4$ or 6, *and in the cases* $q = 3,4,6$ *there is a vector to such that* $T = \{n_1t_o+n_2\theta t_o;\ n_1,n_2 \text{ integers}\}$. First choose t_o and θ as above. Since t_o is a vector of minimum length and $t_o \neq \theta t_o$ the vector $t_o-\theta t_o$ must have length not less than that of t_o. Therefore $q \leq 6$. Similarly $|t_o+\theta^2t_o| \geq |t_o|$ so $q \neq 5$. If $q \geq 2$ and $x_1t_o-x_2\theta t_o \in T$ for non-integral x_1,x_2 then we can choose x_1,x_2 between 0 and $\frac{1}{2}$ so that

$$x_1t_o-x_2\theta t_o = (x_1^2+x_2^2-2x_1x_2\cos\theta)^{\frac{1}{2}}|t_o| \leq (\tfrac{1}{4}+\tfrac{1}{4}-\tfrac{1}{2}\cos\theta)^{\frac{1}{2}}|t_o| \leq |t_o|$$

with equality only if $q = 2$ and $x_1 = x_2 = \frac{1}{2}$. In the cases $q = 3,4,6$ this contradicts the fact that t is of minimum length and so x_1,x_2 must be integral.

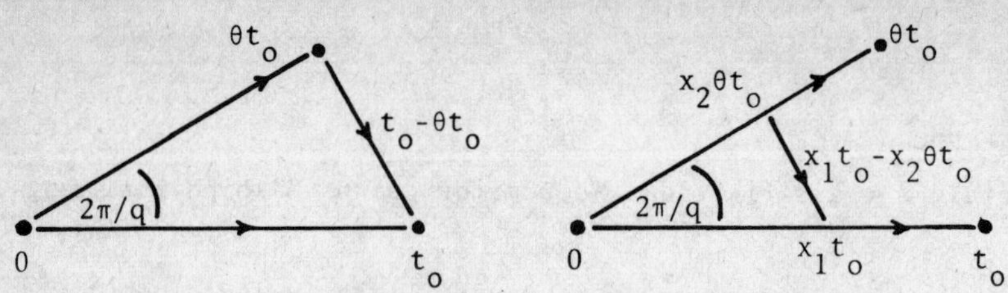

(iv) *shift vectors can be chosen zero when* $q = 3,6$ *and determine the existence of glide symmetries when* $q = 1,2,4$. It is only necessary to consider the case in which ϕ is reflection in some line ℓ. Let $r \in T$ be a non-zero vector on ℓ, and chosen to be of minimum length among such vectors. If $r = t+\phi t$ for some $t \in T$ then, since the shift vector $a \in T$ must lie on ℓ, it is also true that $a=t+\phi t$ for some $t \in T$ and hence that ϕ has a non-zero shift vector. If on the other hand r is not of the form $t+\phi t$ then the shift vectors are determined modulo even multiples of r and can be chosen to be $a = 0$ or $a = r$. If $a = 0$ then $(v,\phi) \in G$ where v is perpendicular to ℓ and therefore represents a reflection (or "mirror symmetry") while if $a = r$ then v has component $\frac{1}{2}r$ parallel to ℓ and therefore represents a "glide symmetry".

1.3. EQUIVALENCE OF PLANE GROUPS

For the purpose of this introductory chapter we shall adopt an apparently restrictive definition of equivalence. Later results will imply that the restriction can be removed (see 2.4 and 3.2).

Definition Two plane groups G,G' with lattices T,T' are *equivalent* if there is an isomorphism $G \to G'$ which maps $T \to T'$.

The consequences of this definition for the invariants (i)-(iv) are

(i) Restriction of the isomorphism to T yields a map $\lambda : T \to T'$ which, since T contains a pair of linearly independent vectors, extends to an invertible linear transformation λ of R^2.

(ii) If under the isomorphism (v,ϕ) maps to (v',ϕ') then $(\phi t,\iota) = (v,\phi)(t,\iota)(v,\phi)^{-1}$ must map to $(v',\phi')(\lambda t,\iota)(v',\phi')^{-1} = (\phi'\lambda t,\iota)$ and therefore $\lambda\phi t = \phi'\lambda t$ for all $t \in T$; since T contains a pair of linearly independent vectors $\lambda\phi = \phi'\lambda$ and $H' = \lambda H\lambda^{-1}$, $H'_o = \lambda H_o\lambda^{-1}$.

(iii) The above maps are compatible with the actions of H on T, and of H' on T'.

(iv) If $\phi \in H$ has order q then so does $\phi' = \lambda\phi\lambda^{-1}$; since $(a,\iota) = (v,\phi)^q$ maps to $(v',\phi')^q = (a',\iota)$ the shift vectors satisfy $a' = \lambda a$.

1.4. CLASSIFICATION THEOREMS

The 17 equivalence classes may be conveniently obtained in three stages: there are 5 for which the point group contains no reflections (i.e. $H_o = H$), 7 further groups for $q = 1,2$, and 5 further groups for the cases $q = 3,4,6$. In each case the method is to show that the invariants (i)-(iv) determine the equivalence class uniquely.

Theorem There are five equivalence classes of plane group G whose point group contains no reflections.

Proof By 1.2 the point group is cyclic with generator θ rotation through $2\pi/q$ where $q = 1,2,3,4$ or 6. Let G and G' be two plane groups with the same value of q, so that the point groups H and H' have the same generator θ. To show that G and G' are equivalent construct an isomorphism $\lambda : T \to T'$ such that $\theta\lambda = \lambda\theta$ (for $q = 1,2$ any isomorphism will do; for $q = 3,4,6$ choose $t_o \in T$, $t'_o \in T'$ as in 1.2(iii) and define $\lambda t_o = t'_o$, $\lambda\theta t_o = \theta t'_o$).
Then $\theta^i\lambda = \lambda\theta^i$ for $i = 0,1,\ldots,q-1$. Each of the groups G,G' is a union of cosets

$$G = T(0,\iota) \cup T(v,\theta) \cup \ldots\ldots \cup T(v,\theta)^{q-1},$$

$$G' = T'(0,\iota) \cup T'(v',\theta) \cup \ldots\ldots \cup T'(v',\theta)^{q-1}.$$

Define $G \to G'$ by mapping $(t,\iota)(v,\theta)^i$ to $(\lambda t,\iota)(v',\theta)^i$. This is a homomorphism because of the relations

$$(v,\theta)^i(t,\iota) = (\theta^i t,\iota)(v,\theta)^i \quad , \quad (v,\theta)^q = (0,\iota),$$

$$(v',\theta)^i(\lambda t,\iota) = (\theta^i \lambda t,\iota)(v'\theta)^i = (\lambda\theta^i t,\iota)(v',\theta)^i, \quad (v',\theta)^q = (0,\iota).$$

Moreover the homomorphism is invertible and maps T on to T'. Therefore G and G' are equivalent as required. The five equivalence classes will be denoted P1, P2, P3, P4, P6, the symbol P (for primitive) referring to the choice of basis for T. Five patterns for which the appropriate G is the group of symmetries are given in the diagram.

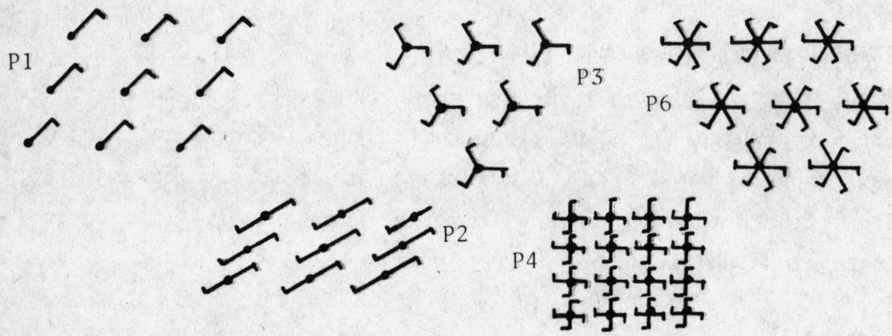

<u>Theorem</u> There are three equivalence classes of plane group G whose point group contains a single reflection.

<u>Proof</u> Let the generator of H be the reflection ρ in the line ℓ. Then $(v,\rho) \in G$ for some $v \in R^2$ which depends on the choice of origin in the plane. We shall suppose that v is chosen to lie on ℓ (if not then replace G by the equivalent group $(-x,\iota)G(x,\iota)$ which contains the element $(-x+v+\rho x,\rho) = (v-2x,\rho)$ for x perpendicular to ℓ). By 1.2(iv) there are three possible configurations of the various invariants. If $a \in T$ is the shift vector of ρ and $r \in T$ is a non-zero vector of minimum length on ℓ then:

$$r = t+\rho t \text{ for some } t \in T \Rightarrow a = 0 \Rightarrow v = 0 \ , \qquad [Cm]$$

$$r \neq t+\rho t \Rightarrow \begin{cases} \text{either } a = 0 \Rightarrow v = 0 \ , & [Pm] \\ \text{or} \quad a = r \Rightarrow v = \tfrac{1}{2}r \ . & [Pg] \end{cases}$$

The symbols in square brackets denote the three configurations:
P = primitive, C = centred, m = mirror, g = glide.

If now G and G' are two plane groups which determine the same configuration, so that $a,r \in T$ lie on ℓ and $a',r' \in T'$ lie on ℓ', there is an isomorphism $\lambda : T \to T'$ defined by $\lambda r = r'$, $\lambda s = s'$. Here $s \in T$, $s' \in T'$ are non-zero vectors perpendicular to ℓ, ℓ' respectively and of minimum length among such vectors: their existence is assured because for any $t \in T$, $t' \in T'$ the vectors $t-\rho t$, $t'-\rho' t'$ are perpendicular to ℓ, ℓ'. Now define $G \to G'$ by mapping $(t,\iota)(v,\rho)^i$ to $(\lambda t,\iota)(v',\rho')^i$ for $i = 0,1$. This is a homomorphism because of the relations

$$(v,\rho)(t,\iota) = (\rho t,\iota)(v,\rho) \quad , \quad (v,\rho)^2 = (a,\iota)$$

$$(v',\rho')(\lambda t,\iota) = (\rho'\lambda t,\iota)(v',\rho') = (\lambda\rho t,\iota)(v',\rho'),(v',\rho')^2 = (a',\iota) = (\lambda a,\iota)$$

and it maps T on to T'. Thus G and G' are equivalent. Three patterns for which the appropriate G is the group of symmetries are given in the diagram.

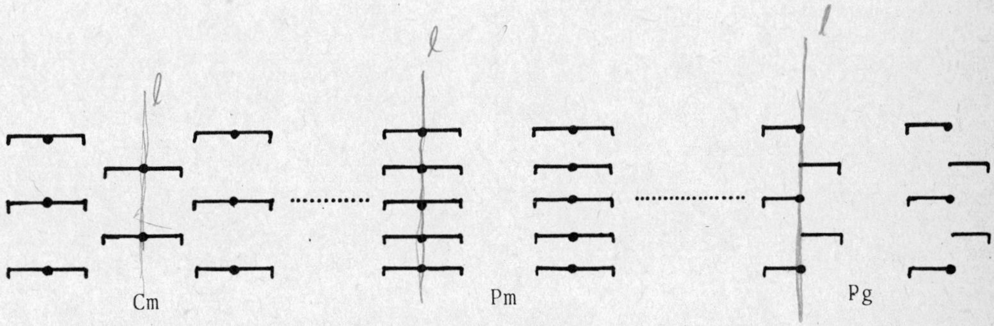

<u>Theorem</u> There are four equivalence classes of plane group G whose point group is generated by two perpendicular reflections.

<u>Proof</u> Let $\rho,\sigma \in H$ be reflections in perpendicular lines ℓ,m. Then $(v,\rho) \in G$, $(w,\sigma) \in G$ for $v,w \in R^2$ which we suppose chosen to lie on ℓ,m respectively (if not then replace G by the equivalent group $(-x-y,\iota)G(x+y,\iota)$ which contains the elements $(v-2x,\rho),(w-2y,\sigma)$ for x on m and y on ℓ). If

either r or s has the form $t+\rho t$ or $t+\sigma t$ then so does the other; in fact then $t = \frac{1}{2}r+\frac{1}{2}s \in T$. There are therefore four possible configurations of the various invariants. If $a,b \in T$ are the shift vectors of ρ,σ, and $r,s \in T$ are as above then:

$$\tfrac{1}{2}r+\tfrac{1}{2}s \in T \implies a=b=0 \implies v=w=0 \qquad \text{[Cmm]}$$

$$\tfrac{1}{2}r+\tfrac{1}{2}s \notin T \implies \begin{cases} a=b=0 \implies v=w=0 & \text{[Pmm]} \\ a=0,b=s \implies v=0,w=\tfrac{1}{2}s & \text{[Pgm]} \\ a=r,b=s \implies v=\tfrac{1}{2}r,w=0 & \text{[Pgg]} \end{cases}$$

Note that if ρ,σ yield one mirror and one glide we choose ℓ to be the mirror line and m to be the glide line. Again given G,G' with the same configuration we may define $G \to G'$ by mapping (t,ι) to $(\lambda t,\iota)$, (v,ρ) to (v',ρ') and (w,σ) to (w',σ'). This is a homomorphism because of the previous relations together with

$$(v,\rho)(w,\sigma) = (w,\sigma)^{-1}(v,\rho)^{-1} ,$$

$$(v',\rho')(w',\sigma') = (w',\sigma')^{-1}(v',\rho')^{-1},$$

and so G and G' are equivalent.

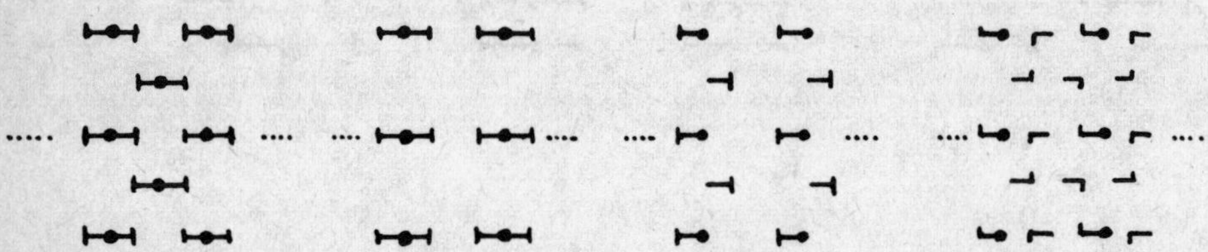

Cmm Pmm Pgm Pgg

It remains to consider the cases in which H contains more than two reflections, i.e. q = 3,4 or 6. Though H is still generated by two reflections, these are now in lines at angle π/q and the above proof must be modified particularly as regards the choice of origin. In all cases we shall choose the origin at a centre of rotation and assume from the outset that $(0,\theta) \in G$ (if not then replace G by the equivalent group $(w,\iota)G(-w,\iota)$

which contains the element $(w,\iota)(v,\theta)(-w,\iota) = (v+w-\theta w,\theta))$.

Theorem There are five equivalence classes of plane group G whose point group contains more than two reflections.

Proof H is generated by the rotation θ through $2\pi/q$ and reflection ρ in a line ℓ. The latter can always be chosen in one of the positions:

$(q=3)$ ℓ through $-t_o$		[P31m]
$(q=3)$ ℓ through $t_o+2\theta t_o$	$= \theta t_o+\rho\theta t_o \implies a=0$	[P3m1]
$(q=6)$ ℓ through t_o		[P6mm]
$(q=4)$ ℓ through t_o	either $a=0$	[P4mm]
	or $a=\frac{1}{2}t_o$	[P4gm]

where $a \in T$ is the shift vector of ρ. Note that in the first three cases $a=0$ by 1.2(iv) and therefore $(v,\rho) \in G$ with v perpendicular to ℓ; that is, (v,ρ) is a reflection in a line parallel to ℓ. Since the intersection of two lines of reflection at angle π/q is always the centre of a rotation through $2\pi/q$ we may choose the origin to be at such a point. Therefore we may in the first four cases assume $(0,\theta) \in G$, $(0,\rho) \in G$. If G and G' are both as in one of the first four cases above we have

$$G = \bigcup_{i=0}^{q-1} T(0,\theta)^i \cup \bigcup_{i=0}^{q-1} T(0,\theta)^i(0,\rho)$$

$$G' = \bigcup_{i=0}^{q-1} T'(0,\theta)^i \cup \bigcup_{i=0}^{q-1} T'(0,\theta)^i(0,\rho')$$

and can construct an isomorphism $G \to G'$ by defining $\lambda t_o=t_o'$, $\lambda\theta t_o = \theta t_o'$ and mapping $(t,\iota)(0,\theta)^i(0,\rho)^j$ to $(\lambda t,\iota)(0,\theta)^i(0,\rho')^j$ for $i = 0,\ldots,q-1$ and $j = 0,1$. There remains the fifth and final case when $q = 4$ and $a = \frac{1}{2}t_o$. We assume that $(0,\theta) \in G$, $(u+\frac{1}{2}t_o,\rho) \in G$ where u is perpendicular to t_o. The latter element is a glide symmetry along a line parallel to ℓ but displaced by $\frac{1}{2}u$; there is a second such glide symmetry along the line parallel to ℓ but displaced by $\theta t_o-\frac{1}{2}u$. The product of these two glide symmetries is

$$(2\theta t_o - u + \tfrac{1}{2}t_o, \rho)(u + \tfrac{1}{2}t_o, \rho) = (2\theta t_o - 2u + t_o, \iota) \in T$$

and therefore u is an odd multiple of $\frac{1}{2}\theta t_o$ which can be taken to be $\frac{1}{2}\theta t_o$ itself. We now have

$$G = \bigcup_{i=0}^{4} T(0,\theta)^i \cup \bigcup_{i=0}^{4} T(0,\theta)^i(\tfrac{1}{2}t_o + \tfrac{1}{2}\theta t_o, \rho)$$

$$G' = \bigcup_{i=0}^{4} T'(0,\theta)^i \cup \bigcup_{i=0}^{4} T'(0,\theta)^i(\tfrac{1}{2}t'_o + \tfrac{1}{2}\theta t'_o, \rho')$$

and define $\lambda t_o = t'_o$, $\lambda\theta t_o = \theta t'_o$ as before. We conclude that there are five equivalence classes denoted P31m, P3m1, P6mm, P4mm, P4gm corresponding to symmetry groups of patterns like those in the diagram.

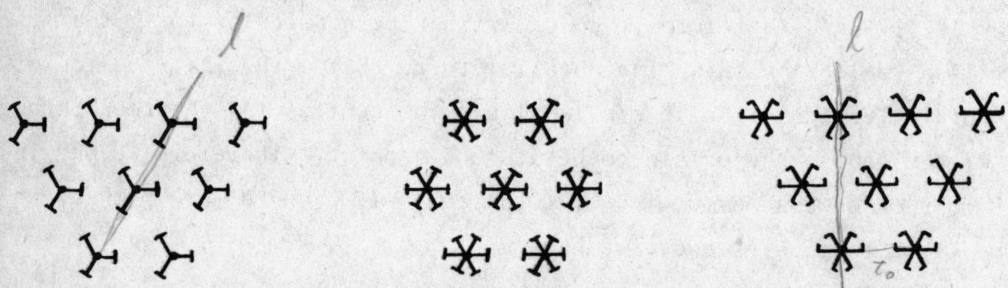

P31m P6mm P3m1

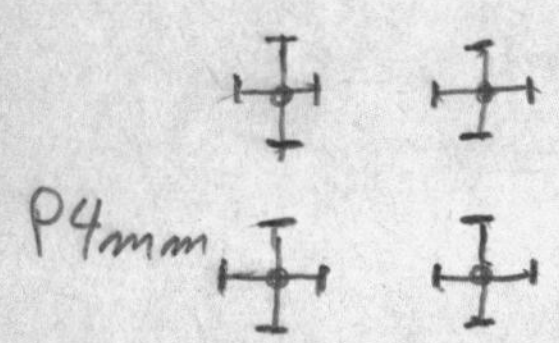

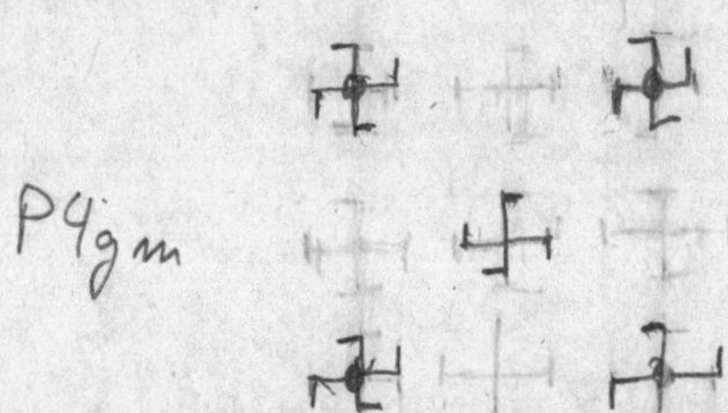

1.5. DEFORMATIONS OF LATTICES

It will be noticed that the 17 equivalence classes of plane groups yield a much smaller number of lattices T. Let G(T) be the group of all rotations and reflections which map T to itself. There are *four* distinct possibilities for G(T); in the table G(T) is listed according to its most convenient generators

G	G(T)
P1, P2	2
P3, P6, P31m, P3m1	6m (=6mm)
Pm, Pg, Pmm, Pgm, Pgg	mm (=2mm)
Cm, Cmm	mm (=2mm)
P4, P4mm, P4gm	4m (=4mm)

If on the other hand the pairs (T,G(T)) are classified up to the natural definition of equivalence arising from 1.3, namely there is an isomorphism $\lambda : T \to T'$ such that $G(T') = \lambda G(T)\lambda^{-1}$, then there are *five* equivalence classes because the primitive lattice P is now not equivalent to the centred lattice C (Proof: P has an integral basis consisting of vectors r,s on lines of reflection, whereas C does not because it contains the vector $\frac{1}{2}r+\frac{1}{2}s$). Finally there is a more topological classification which arranges the possible lattices in *six* equivalence classes as follows.

Let $t_o \in T$ be a non-zero vector of minimum length. Since T admits a pair of integral generators one of which is t_o, it is possible to plot in R^2 the possible positions of the second generator t_1 relative to t_o. Replacing t_o by $-t_o$, and t_1 by $-t_1$ we may assume that t_1 lies in the first quadrant relative to t_o.

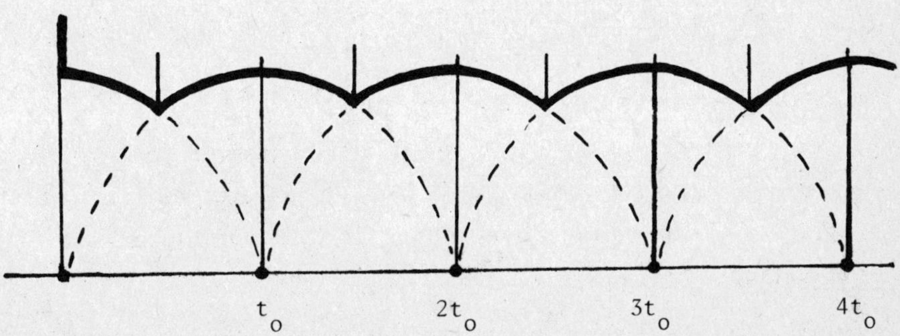

Clearly t_1 lies within the region bounded by the thick line, otherwise t_o would not be a non-zero vector of minimum length. The repetition of regions bounded by vertical lines corresponds to successive application of the maps $t_1 \mapsto t_1+t_o$ and $t_1 \mapsto -t_i+t_o$. A single region has the form

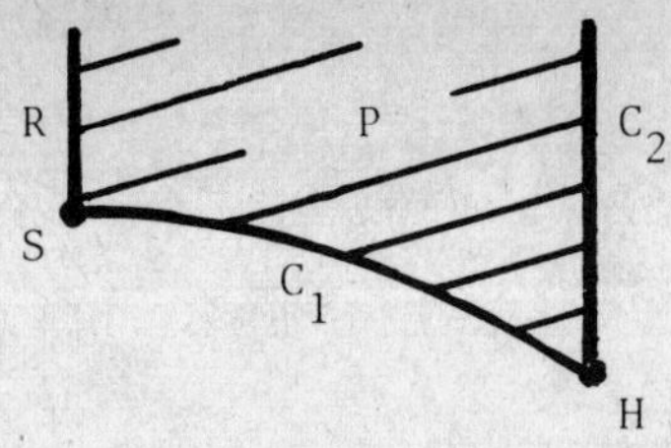

where now the various parts of the region correspond to qualitatively different positions for t_1. If (,) denotes scalar products of vectors then the relationship between parts of the diagram and lattices is

P - parallelogram	$0 < 2(t_o,t_1) < (t_o,t_o) < (t_1,t_1)$
R - rectangle	$0 = 2(t_o,t_1) < (t_o,t_o) < (t_1,t_1)$
C - centred rectangle or diamond	$(C_1)\ 0 < 2(t_o,t_1) < (t_o,t_o) = (t_1,t_1)$
	$(C_2)\ 0 < 2(t_o,t_1) = (t_o,t_o) < (t_1,t_1)$
H - hexagon or equilateral triangles	$0 < 2(t_o,t_1) = (t_o,t_o) = (t_1,t_1)$
S - square	$0 = 2(t_o,t_1) < (t_o,t_o) = (t_1,t_1)$

for suitable choice of t_1. Two lattices T,T' determine the same point on this diagram if and only if there is a scalar multiple c and a rotation or reflection θ such that $cT' = \theta T$.

1.6. STRATEGY FOR CLASSIFICATION IN HIGHER DIMENSIONS

It is worth making three comments about the results obtained in 1.4. Firstly, the definition of equivalence in 1.3 was chosen to be intermediate between stronger and weaker definitions. Thus if G,G' are plane groups there are implications

$G' = (v,\phi)G(v,\phi)^{-1}$ with $v \in R^2$ and ϕ linear

$\Longrightarrow$

there is an isomorphism $G \to G'$ which maps $T \to T'$

$\Longrightarrow$

there is an isomorphism $G \to G'$.

Inspection of the 17 equivalence classes shows that they are unchanged whichever definition is adopted. It will be shown in 2.4 and 3.2 that this is always the case: the above three statements are identical provided that G and G' satisfy the n-dimensional analogue of the definition of plane group.

Secondly it will be noted that the discussion of possible lattices in 1.5 is independent of that in 1.2-1.4. There are in fact two quite distinct ways of approaching the classification of space groups. In both we consider possible lattices $T \subset R^n$ and possible finite subgroups H of the group O(n) of all linear transformations which preserve distance and angle. The first way, followed in 1.2-1.4, is to begin by listing all H which can act on some lattice of rank n (this stage corresponds to the *crystallographic restriction* q = 1,2,3,4 or 6 in 1.2(iii)) and then to ask, for each H in turn, what are the particular T on which H can act, and what are the possible positions of the shift vectors. The second way is to begin by analysing the various possibilities for T, as in 1.5, and then to ask, for each type of lattice T in turn, what are the subgroups H of the symmetry group $G(T) = \{\theta \in O(n);\ \theta T = T\}$ of T, and what are the possible positions of the shift vectors.

Thirdly it might be expected that an inductive procedure might relate symmetry groups in dimension n to more complicated groups arising from dimension n-1. For example, the two lattices P,C with symmetry group mm are closely related to the 1-dimensional objects T_p, T_c illustrated.

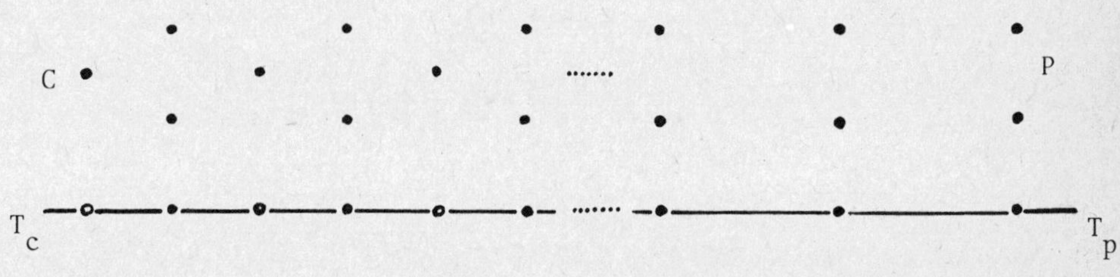

while the two plane groups denoted Pm and Pg are closely related to the 1-dimensional symmetry groups of the horizontal pattern illustrated:

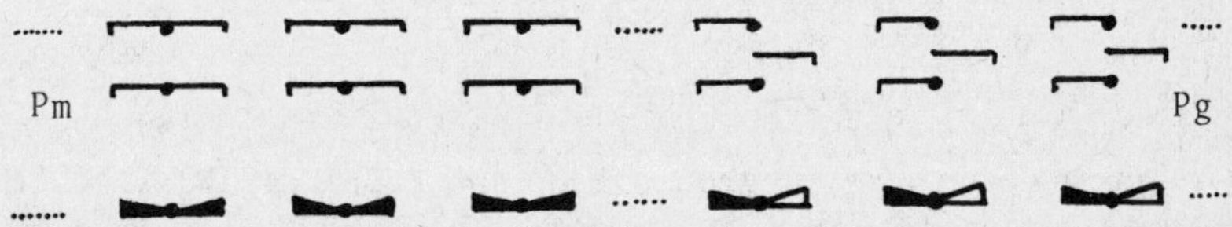

In each case the second 1-dimensional pattern yields a *black and white* symmetry group in which some symmetries reverse colouring and some do not. Thus a study of *black and white* groups in the plane, apart from being of interest in its own right, will give information about certain classes of symmetry groups in 3-dimensional space. The classification of *black and white* groups in the plane is an easy deduction from that of plane groups and is carried out in the next section.

1.7. BLACK AND WHITE GROUPS

Definition A black and white group is a pair $G_1 \subset G$ of plane groups such that G_1 is a subgroup of G of index 2. Two black and white groups $G_1 \subset G$ and $G_1' \subset G'$ are equivalent if there is an isomorphism $G \to G'$ which maps G_1 on to G_1' and T on to T'.

Remarks (i) It is a consequence of the definition of equivalence that the isomorphism maps $T_1 = G_1 \cap T$ on $T_1' = G_1' \cap T'$ and so induces an equivalence between G_1 and G_1'.
(ii) We think of G_1 as the subgroup which preserves the two colours, and of the complement $G \backslash G_1$ as the coset which interchanges the two colours. Some authors allow the latter set to be empty, or even to overlap with G_1 (producing so-called *grey* groups) but this does not lead to any genuinely new constructions.

There are two quite distinct ways in which black and white groups can arise. Let T_1, T and H_1, H denote the lattice and point group of G_1, G respectively. Since G_1 is of index 2 in G it follows that

EITHER $T_1 = T$ and $H_1 \subset H$ is a subgroup of index 2

OR $T_1 \subset T$ is a subgroup of index 2 and $H_1 = H$.

Theorem There are 26 equivalence classes of black and white groups with $T_1 = T$ (the *zellengleich* or *translation equivalent* case).

Proof It is sufficient to go through the list of 17 equivalence classes of plane group, asking which point groups have subgroups of index 2. The first two theorems of 1.4 yield 6 black and white groups as indicated:

G	P1	P2	P3	P4	P6	Cm	Pm	Pg
H	1	2	3	4	6	m	m	m
H_1	-	-	-	2	3	1	1	1
G_1	-	P1	-	P2	P3	P1	P1	P1

The third theorem of 1.4 yields 9 groups (here m_1, m_2 refer to mirrors along first or second coordinate directions);

G	Cmm		Pmm		Pgm			Pgg	
H	2mm		2mm		2mm			2mm	
H_1	2	m	2	m	2	m_1	m_2	2	m
G_1	P2	Cm	P2	Pm	P2	Pm	Pg	P2	Pg

Finally the fourth theorem of 1.4 yields 11 groups (here m_1, m_2 refer to reflections which are, are not in lines through a non-zero vector of minimum length):

G	P31m	P3m1	P6mm			P4mm			P4gm		
H	3m	3m	6mm			4mm			4mm		
H_1	3	3	6	31m	3m1	4	$2m_1$	$2m_2$	4	$2m_1$	$2m_2$
G_1	P3	P3	P6	P31m	P3m1	P4	Pmm	Cmm	P4	Pgg	Cmm

There should be no difficulty in drawing black and white patterns appropriate to each of the 26 pairs $G_1 \subset G$. The last four pairs are of particular interest:

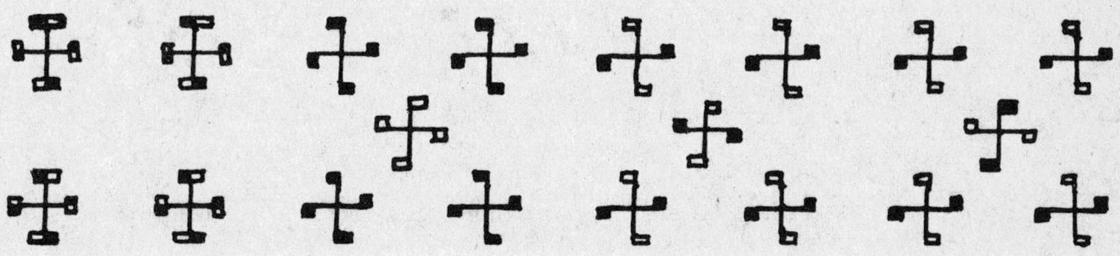

Theorem There are 20 equivalence classes of black and white groups with $H_1 = H$ (the *klassengleich* or *class equivalent* case).

Proof It is sufficient to go through the list of 17 equivalence classes of plane group, asking which lattices have sublattices of index 2 invariant under the action of H. It is easy to see that this is impossible for H = 3,6,3m,6m. The remaining point groups yield the sublattices indicated:

H = 1,2

H = m,mm

H = 4,4m

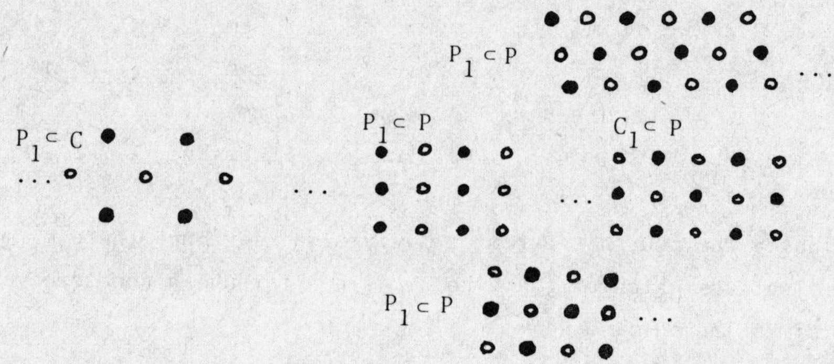

Such sublattices may or may not be consistent with a given plane group since for each $\theta \in H$ it must be possible to find $v \in R^2$ such that $(v,\theta) \in G_1$ with shift vector $a \in T_1$. Moreover two choices v and v+t, with $t \in T$, may now yield different subgroups $G_1 \subset G$. If H contains no reflections such problems do not arise: we obtain 3 black and white groups $P_1 1 \subset P1$, $P_1 2 \subset P2$, $P_1 4 \subset P4$ which are illustrated. If H contains reflections with

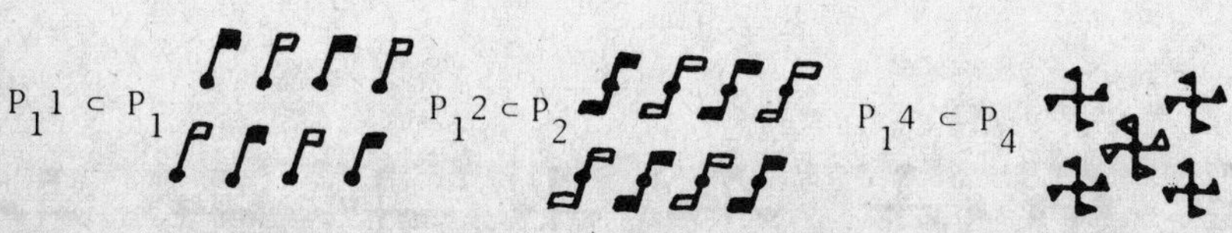

shift vector non-zero, so that G containsglide symmetries, then G_1 must also contain glide symmetries (because $(v,\rho) \notin G_1$ implies $(v+t,\rho) \in G_1$ for $t \in T \backslash T_1$). If however H contains reflections ρ with shift vector zero then G_1 contains elements of the form (t,ρ), $t \in T$ and there are two cases according as $t \in T_1$ or $t \notin T_1$. We obtain the 17 black and white groups listed.

G	G_1
Pm	$P_{11}m$, $P_{12}m$, P_1g, C_1m
Pg	P_1g
Cm	P_1m, P_1g
Pmm	P_1mm, $P_{11}mg$, C_1mm
Pgm	$P_{12}gm$, $P_{11}gg$
Cmm	P_1mm, $P_{11}mg$, P_1gg
P4mm	P_14mm, P_14gm

Here P_{11}, P_{12} refer to whether first or second axis contains black and white vectors, while a symbol g for G_1 replacing m for G signals the case $t \notin T_1$ mentioned above.

$P_1g \subset Cm$ 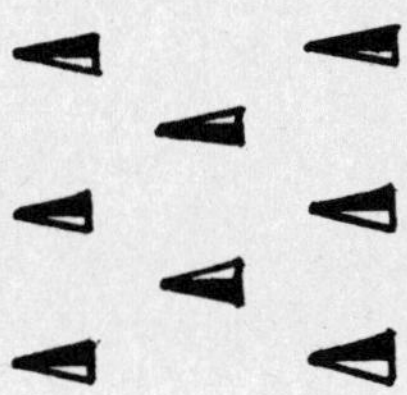

$P_14gm \subset P4mm$ 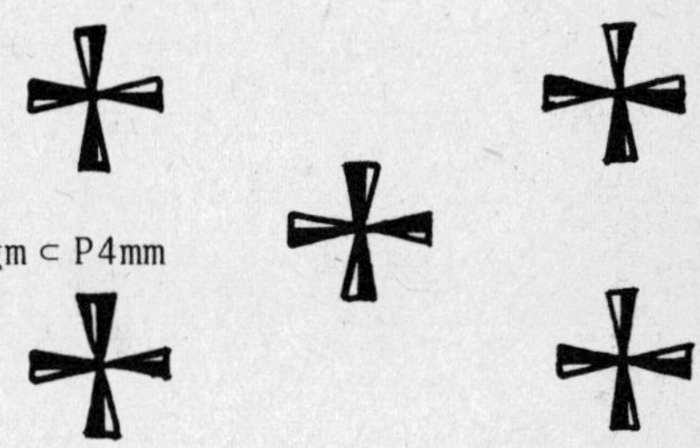

2 Groups of affine transformations

The space in which we live is 3-dimensional but does not have a preferred origin. Properties which do not depend on a choice of origin, for example the centre of gravity of a system of particles, are therefore of physical significance. We shall develop these properties for n-dimensional space.

2.1. THE CANONICAL EXACT SEQUENCE

Let V be a vector space with origin $v_o \in V$. If $v_1 \in V$ is another possible choice of origin then a new scalar multiplication $\cdot$ and addition $\dotplus$ may be defined by

$$\lambda\cdot v \dotplus \mu\cdot w = \lambda \underset{\bullet}{v} + \mu \underset{\bullet}{w} + (1-\lambda-\mu)v_1 \quad .$$

It is easy to check that the axioms for a vector space are satisfied with v_1 as origin and with operations $\cdot$ and $\dotplus$. Moreover if $\lambda_1+\ldots+\lambda_r = 1$ then the sum $\lambda_1u_1+\ldots+\lambda_ru_r$ does not depend on choice of origin at all. It is possible to define the notion of *abstract* affine space by postulating the existence of a sum $\lambda_1u_1+\ldots+\lambda_ru_r$ whenever $\lambda_1+\ldots+\lambda_r = 1$, but we proceed directly to consider the group of invertible affine transformations which preserve such a structure.

Theorem The set A of invertible affine transformations forms a group under composition. The subset T of those of the form $u \longmapsto u+v_o-u_o$ is an abelian normal subgroup of A.

Proof It is necessary to show that f^{-1} is an affine transformation for each $f \in A$. If $u_i = f(v_i)$ then $v_i = f^{-1}(u_i)$ and we may write $f^{-1}(\lambda_1u_1+\ldots+\lambda_ru_r)$ in the form

$$f^{-1}(\lambda_1f(v_1)+..+\lambda_rf(v_r)) = f^{-1}f(\lambda_1f_1+\ldots+\lambda_rv_r) =$$
$$= \lambda_1v_1+\ldots+\lambda_rv_r = \lambda_1f^{-1}(u_1)+\ldots+\lambda_rf^{-1}(u_r).$$

It follows that A is a group. Consider now two elements of T denoted $u \mapsto u+v_o-u_o$ and $u \mapsto u+v_1-u_1$. Their composite is

$$u \mapsto u+v_o-u_o+v_1-u_1 = u+(v_o+v_1-w)-(u_o+u_1-w)$$

for arbitrary w. Therefore T is an abelian subgroup of A. Finally if $f \in A$ then $f\tau f^{-1}$ maps to

$$f(f^{-1}(u)+v_o-u_o) = u+f(v_o)-f(u_o)$$

and so $f\tau f^{-1} \in T$.

Definition The quotient group $L = A/T$ is called the *linear group*. The action of L on T by conjugation is called the *canonical action* and the corresponding exact sequence

$$1 \longrightarrow T \longrightarrow A \longrightarrow L \longrightarrow 1$$

is called the *canonical exact sequence*.

The name chosen for L is justified by

Theorem T admits the structure of a vector space and the action of L on T is then linear.

Proof The essential point is that T has a preferred origin namely the transformation $u \mapsto u+w-w$ for arbitrary w. Since T is abelian we define h+k by composition. Thus

$$h+k \text{ maps } u \mapsto u+h(u_o)-u_o+k(u_o)-u_o$$
$$\lambda h \text{ maps } u \mapsto u+\lambda h(u_o)-u_o$$

both descriptions being independent of the choice of u_o because $h(u_o)-u_o = h(u_1)-u_1$ for $h \in T$.

Now let $\phi = fT$ be a coset. It acts on T taking $h \in T$ to the composite fhf^{-1}. Therefore $\phi \cdot h$ is the mapping

$$u \mapsto f(f^{-1}u+h(u_o)-u_o) = u+fh(u_o)-f(u_o)$$

and $\lambda\phi\cdot h+\mu\phi\cdot k$ is the mapping

$$u \mapsto u+f(\lambda h(u_o)+\mu k(u_o)+(1-\lambda-\mu)u_o)-f(u_o)$$

$$= f(f^{-1}u+(\lambda h+\mu k)(u_o)-u_o)$$

which coincides with $\phi(\lambda h+\mu k)$. Thus ϕ acts linearly.

2.2. EFFECT OF CHOICE OF ORIGIN

The results of 2.1 are independent of any choice of origin for V. Once an origin $u_o \in V$ is chosen we have

Theorem There is an identification between T and V under which $0 \in T$ corresponds to $u_o \in V$, and there is a semi-splitting $s : L \to A$ of the canonical exact sequence under which linear transformations $\phi \in L$ correspond to affine transformations which leave u_o fixed.

Proof Map $h \in T$ to $h(u_o) \in V$. The element $v \in V$ arises from a unique element $u \mapsto u+v-u_o$ of T. Similarly map $\phi \in L$ to the affine transformation $u \mapsto f(u)-f(u_o)+u_o$ where $\phi = fT$. The image $s(\phi)$ is well defined because if $k \in T$ then

$$fk(u)-fk(u_o)+u_o = f(u+k(u_o)-u_o)-fk(u_o)+u_o = f(u)-f(u_o)+u_o$$

and s is a homomorphism because it maps the coset fgT to the affine transformation

$$u \mapsto fg(u)-fg(u_o)+u_o = f(g(u)-g(u_o)+u_o)-f(u_o)+u_o.$$

Finally s is a semi-splitting because the transformation $u \mapsto f(u)-f(u_o)+u_o$, being the composite of the transformation $u \mapsto u-u_o+f^{-1}(u_o)$ and f itself, lies in the coset fT.

Corollary Once the origin $u_o \in V$ is chosen, every $f \in A$ can be written in a unique way as a pair $f = (h, \ )$ where $h \in T$ and $\phi \in L$.

Proof Given f let $\phi = fT$ and let h be the transformation $u \longmapsto u+f(u_o)-u_o$. Both are determined uniquely by f and u_o. Moreover f is the composite of $s\phi$ and h.

Remarks (i) The corollary implies that f may be written in the form $x \longmapsto h+\phi \cdot x$ where now V is identified with T as in the theorem. We call elements $h \in T$ *translations* and elements $\phi \in L$ *linear transformations*.

(ii) As a set there is a bijection $A \longrightarrow T \times L$ but this is not consistent with the group multiplication because the product of $f = (h,\phi)$ and $g = (k,\psi)$ is not $(h+k,\phi\psi)$ but

$$fg = hs(\phi)ks(\psi) = h(s(\phi)ks(\phi)^{-1})s(\phi)s(\psi) = h(\phi\cdot k)s(\phi\psi) = (h+\phi k,\phi\mu).$$

2.3. GROUPS OF ISOMETRIES AND OTHER SUBGROUPS

Consider the effect of having distance defined on V. We denote the distance between $u,v \in V$ by $|u-v|$. It does not depend on choice of origin. However, once an origin $0 \in V$ is given, it gives modulus $|u| = |u-0|$ and scalar product $(u,v)=\frac{1}{2}(|u+v|^2-|u|^2 -|v|^2)$.

Definition A function $f : V \to V$ is an *isometry* if $|f(u)-f(v)| = |u-v|$ for all $u,v \in V$.

Theorem If f is an isometry then f is an affine transformation.

Proof Suppose an origin $0 \in V$ is given and $w = f(0)$. Then the function ϕ defined by $u \longmapsto f(u)-w$ keeps 0 fixed and $f(x) = w+\phi(x)$. We show that ϕ is linear. By the formula for scalar product $\{e_1,\dots e_n\}$ orthonormal basis implies $\{\phi(e_1),\dots,\phi(e_n)\}$ orthonormal basis; moreover if $u = \alpha_1 e_1+\dots\alpha_n e_n$ then $\alpha_i = (u,e_i)$ and hence $\phi(u) = \alpha_1\phi(e_1)+\dots+\alpha_n\phi(e_n)$.

Thus the isometries form a subgroup G of A. Those which keep the origin fixed can be identified, as in 2.2, with a subgroup O of L. Every isometry f can be written in the form (w,ϕ) with $w \in T$ and $\phi \in O$. The groups G and

$\mathcal{O}$ are called the *Galilean group* and the *orthogonal group* respectively. The situation can be summarised by the exact sequences

$$\begin{array}{ccccccccc} 1 & \longrightarrow & \mathcal{T} & \longrightarrow & \mathcal{A} & \xrightarrow{p} & \mathcal{L} & \longrightarrow & 1 \\ & & \| & & \cup & & \cup & & \\ 1 & \longrightarrow & \mathcal{T} & \longrightarrow & \mathcal{G} & \longrightarrow & \mathcal{O} & \longrightarrow & 1 \end{array}$$

where unbroken arrows denote homomorphisms defined canonically and broken arrows denote homomorphisms which depend on a choice of origin. It is an example of what happens whenever a subgroup $G \subset \mathcal{A}$ is specified.

Definition Let $G \subset \mathcal{A}$ be a subgroup. The *lattice* T of G is the intersection $T = \mathcal{T} \cap G$. The *point group* H of G is the image $p(G) \subset \mathcal{L}$.

Thus we always have a diagram of exact sequences

$$\begin{array}{ccccccccc} 1 & \longrightarrow & \mathcal{T} & \longrightarrow & \mathcal{A} & \xrightarrow{p} & \mathcal{L} & \longrightarrow & 1 \\ & & \cup & & \cup & & \cup & & \\ 0 & \longrightarrow & T & \longrightarrow & G & \longrightarrow & H & \longrightarrow & 1 \end{array}$$

The use of 0 instead of 1 in the second exact sequence is because in practice T is usually written additivity, with composition defined by addition of vectors rather than by composition of translations.

Remarks (i) As before, every element γ of G can be written in the form $\gamma = (h,\phi)$ with $h \in \mathcal{T}$, $\phi \in \mathcal{L}$. It follows immediately that $\phi \in H$.

(ii) It is *not* true that h can be chosen in T for all G. Examples arise already when V is a plane and G contains glide transformations.

(iii) There is always an action of H on T by conjugation $(h,\phi)(t,\iota)(h,\phi)^{-1} = (\phi t,\iota)$.

The lattice T should be thought of as a set of translation vectors rather than as a set of points of V. Of course once an origin is chosen the two are identified, but T is independent of such a choice. Indeed T has a canonical origin, namely the zero translation defined by the identity map id_V.

Definition Two subgroups G_1, G_2 of A are *equivalent* if $G_2 = fG_1f^{-1}$ for some $f \in A$.

Consider the consequences of this if $f = (k, \psi)$. There is a diagram of exact sequences

$$\begin{array}{ccccccccc}
0 & \longrightarrow & T_1 & \longrightarrow & G_1 & \longrightarrow & H_1 & \longrightarrow & 1 \\
 & & \downarrow{\scriptstyle \phi} & & \downarrow & & \downarrow & & \\
0 & \longrightarrow & T_2 & \longrightarrow & G_2 & \longrightarrow & H_2 & \longrightarrow & 1
\end{array}$$

in which vertical arrows denote the homomorphisms (from left to right) ψ, conjugation by (k, ψ) conjugation by ψ. Here we regard T_1 and T_2 as subsets of the vector space T on which ψ acts linearly.

Theorem If G_1, G_2 are equivalent then there is a linear transformation ψ such that (i) $T_2 = \psi T_1$, (ii) $H_2 = \psi H_1 \psi^{-1}$, (iii) the action of H_2 on T_2 is induced by that of H_1 on T_1. If moreover G_1, G_2 are groups of isometries then there is an orthogonal transformation θ such that $H_2 = \theta H_1 \theta^{-1}$.

Proof The first part follows from the above diagram of exact sequences. To prove the second part it is necessary to show that if $H_2 = \psi H_1 \psi^{-1}$ for $\psi \in L$ then $H_2 = \theta H_1 \theta^{-1}$ for some $\theta \in O$. If $\phi \in H_1$ then $\phi'\phi = \iota$ and $\psi\phi\psi^{-1} \in H_2$. Therefore $\iota = (\psi\phi\psi^{-1})'(\psi\phi\psi^{-1}) = \psi^{-1\prime}\phi'\psi'\psi\phi\psi^{-1}$ and hence

$$\psi^{-1}\psi^{-1\prime}\phi^{-1}\psi'\psi\phi = \iota.$$

In other words ϕ commutes with $\omega = \psi'\psi$. Now ω is positive definite symmetric and so (see Appendix §2) there is $\alpha \in L$ such that $\alpha^2 = \omega$ and $\alpha = a_o + a_1\omega + \ldots + a_m\omega^m$ for some m. The latter relationship implies that α commutes with each element ϕ of H_1. Define $\theta = \psi\alpha^{-1}$. Then

$$\theta'\theta = \alpha^{-1\prime}\psi'\psi\alpha^{-1} = \alpha^{-1\prime}\omega\alpha^{-1} = \alpha^{-1\prime}\alpha^2\alpha^{-1} = \iota$$

and so $\theta \in O$, while

$$\theta\phi\theta^{-1} = \psi\alpha^{-1}\phi\alpha\psi^{-1} = \psi\phi\alpha^{-1}\alpha\psi^{-1} = \psi\phi\psi^{-1}$$

and so $\theta H_1\theta^{-1} = H_2$.

<u>Remark</u> The above proof is the same argument which is used to establish that every $\psi \in L$ can be written in a unique way as a pair $\psi = \theta\alpha$ where $\theta \in O$ and where α is positive definite symmetric.

2.4. GROUPS WITH LARGE TRANSLATION SUBGROUPS

For particular classes of subgroups $G \subset A$ the definition of equivalence can be replaced by one which is apparently weaker.

<u>Definition</u> G has a *large translation subgroup* if (i) the lattice T of G includes a set which spans the vector space T, and (ii) the point group H of G is finite.

<u>Theorem</u> Let $G_1, G_2 \subset A$ have large translation subgroups. Suppose that there is an isomorphism $G_1 \to G_2$ which maps the lattice T_1 of G_1 on to the lattice T_2 of G_2. Then G_1, G_2 are equivalent.

<u>Proof</u> The restriction of the isomorphism $G_1 \to G_2$ defines an isomorphism $\psi : T_1 \to T_2$. Since T_1 spans T, this must be the restriction of a linear transformation $\psi \in L$. Now suppose that the original isomorphism maps $(u_1,\phi_1) \in G_1$ to $(u_2,\phi_2) \in G_2$. Then it must map

$$(\phi_1 t_1,\iota) = (u_1,\phi_1)(t_1,\iota)(u_1,\phi_1)^{-1}$$

to $(u_2,\phi_2)(\psi t_1,\iota)(u_2,\phi_2)^{-1} = (\phi_2\psi t_1,\iota)$. Because T_1 spans T, the relationship $\psi\phi_1 t_1 = \phi_2\psi t_1$ for $t_1 \in T_1$ implies $\phi_2 = \psi\phi_1\psi^{-1}$. Therefore $H_2 = \psi H_1\psi^{-1}$. We may write $u_2 = \psi u_2 + s_{\phi_1}$ where $s_\iota = 0$ and $s_{\alpha\beta} = s_\alpha + \psi\alpha\psi^{-1}s_\beta$. Thus $\{s_\alpha\}_{\alpha\in H_1}$ defines a cocycle $s \in C^1(H_1,T)$ for the action of H_1 on T given by $t \longmapsto \psi\alpha\psi^{-1}t$. By the divisibility theorem (see Appendix §4) this cocycle is a coboundary, i.e. there exists k such that $s_\alpha = k - \psi\alpha\psi^{-1}k$. Therefore (u_2,ϕ_2) must equal

$$(\psi u_1 + k - \psi\alpha\psi^{-1}k,\ \psi\phi_1\psi^{-1}) = (k,\psi)(u_1,\phi_1)(k,\psi)^{-1}$$

and the isomorphism $G_1 \to G_2$ must be induced by conjugation by $f = (k,\psi)$.

Remarks (i) The proof actually shows that the fact that G_2 has large translation subgroup is a consequence of the other hypotheses.

(ii) To see how precisely the finiteness of H_1 is used it is worth writing out the proof of the divisibility theorem in this special case. Sum the equations $s_{\alpha\beta} = s_\alpha + \psi\alpha\psi^{-1}s_\beta$ over all $\beta \in H_1$ for fixed $\alpha \in H_1$, to obtain

$$k' = Ns_\alpha + \psi\alpha\psi^{-1}k'$$

where $k' = \sum_{\beta\in H_1} s_\beta$ and $N = |H_1|$. Now let $k = k'/N$.

(iii) If H_1 is finite it is always possible to modify the distance function so that H_1 becomes a subgroup of $\mathcal{O}$ and hence G_1 a subgroup of $\mathcal{G}$. Namely let the new scalar product of v_1, v_2 be $\frac{1}{N}\sum_{\alpha\in H_1} (v_1, v_2)$. Because of this fact we restrict attention to subgroups $G \subset \mathcal{G}$.

3 Space groups

3.1. DEFINITION AND TOPOLOGICAL PROPERTIES

Definition A subgroup G of G is a *space group* if there is a basis $\{t_1,\ldots,t_n\}$ for the vector space T which is also an integral basis for the lattice T of G.

Here n = dim T is the dimension of the ambient space V. So the definition can also be expressed: G is a group of n-dimensional isometries whose subgroup of translations is generated by a linearly independent set $\{t_1,\ldots,t_n\}$.

Proposition If G is a space group then there is a non-zero translation $t \in T$ of minimum length and the point group H is finite.

Proof An integral combination $t = \sum_{i=1}^{n} r_i t_i$ has length $|t| < c$ if

$$\sum_{i,j=1}^{n} r_i r_j (t_i, t_j) < c^2.$$

The quadratic form $Q(x) = \sum_{i,j=1}^{n} x_i x_j (t_i, t_j)$ is positive definite because it represents the length of a vector in $T = R^n$. Solutions of the inequality $Q(x) < c^2$ lie in a convex bounded region of R^n. But such a region contains only a finite number of integer points, so that the number of integral solutions is finite. It follows that there is a non-zero solution (not unique) of minimum length. Now choose c large enough so that $|t_i| < c$ for $i = 1,\ldots,n$. Then every $\theta \in O = O(n)$ such that $\theta T = T$ determines a permutation of the set $\{t \in T;\ |t| < c\}$ and hence an embedding of the symmetry group

$$G(T) = \{\theta \in O(n);\ \theta T = T\}$$

into a finite group. Since H is a subgroup of G(T) it is also finite.

The above proposition shows that G has large translation subgroup so that the results of 2.4 apply. It also leads to the following topological properties:

<u>Theorem</u> If G is a space group then (i) G is a discrete subgroup of $\mathcal{G}$, (ii) the quotient V/G is compact, and (iii) the quotient $\mathcal{G}$/G is compact.

<u>Proof</u> By the proposition G is a finite union of discrete cosets $T(v_i, \theta_i)$ of T and is therefore discrete. Let P be the closed parallelopiped $\{v = \sum_{i=1}^{n} x_i t_i ;\ 0 \le x_i \le 1\}$. Then P is bounded and hence compact. Similarly $K = P \times O(n)$ is compact (see Appendix §3). There are continuous onto maps

$$\begin{array}{ccccc} P & \longrightarrow & V/T & \longrightarrow & V/G \\ v & \longmapsto & \text{orbit } [v] \text{ under } T & \longmapsto & \text{orbit } [v] \text{ under } G \end{array}$$

$$\begin{array}{ccccc} K & \longrightarrow & \mathcal{G}/T & \longrightarrow & \mathcal{G}/G \\ (v,\theta) & \longmapsto & \text{coset } [(v,\theta)] \text{ of } T & \longmapsto & \text{coset } [(v,\theta)] \text{ of } G \end{array}$$

and therefore the images V/G and $\mathcal{G}$/G are compact.

The above theorem has a partial converse due to Bieberbach which leads to an alternative definition of space groups. We need a number of lemmas on discrete subgroups D of R^n which will then be applied to the lattice T.

<u>Lemma 1</u> Let D be a discrete subgroup of R^n, and $\{d_1,\ldots,d_k\}$ a finite subset of D. If there exist real numbers $x_1,\ldots,x_k$ not all zero for which $\sum_{i=1}^{k} x_i d_i = 0$ then there exist integers $p_1,\ldots,p_k$ not all zero for which $\sum_{i=1}^{k} p_i d_i = 0$.

<u>Proof</u> It is known from number theory that given $\delta > 0$ there is an approximation $p_1/q,\ldots,p_k/q$ to $x_1,\ldots,x_k$ such that

$$|qx_i - p_i| < \delta, \qquad i=1,\ldots,k.$$

Then $|q\Sigma x_i d_i - \Sigma p_i d_i| < \delta\Sigma|d_i|$ and therefore $|\Sigma p_i d_i| < \delta\Sigma|d_i|$. Since D is discrete the left hand side must be zero for sufficiently small choice of δ.

Lemma 2 Let D be a discrete subgroup of R^n. Then D is finitely generated.

Proof Suppose that the subspace of R^n spanned by D has dimension r, and that $\{d_1,\ldots,d_r\}$ is a linearly independent set in D. The parallelopiped $A = \{v = \sum_{i=1}^{r} x_i d_i;\ 0 \le x_i \le 1\}$ is compact and $A \cap D$ generates D as an abelian group. Since D is discrete $A \cap D$ is finite.

Lemma 3 Let D be a discrete subgroup of R^n which spans a subspace of dimension r. Then there is an integral basis $\{d_1,\ldots,d_r\}$ of D.

Proof By Lemma 2 there is an integral basis $\{d_1,\ldots,d_k\}$ for some $k \ge r$ which we suppose is lowest possible. If $k > r$ then by Lemma 1 there exist integers $p_1,\ldots,p_k$ not all zero for which $\sum_{i=1}^{k} p_i d_i = 0$. We may suppose the vectors d_i ordered in such a way that $p_k \ne 0$ and p_1 is the smallest positive p_i for $i = 1,\ldots,k$. Choose m_i so that

$$0 \le p_i - m_i p_1 < p_1, \qquad i = 2,\ldots k-1$$

and replace d_1 by $d_1^{(1)} = d_1+m_2d_2+\ldots+m_{k-1}d_{k-1}$. Then

$$p_1 d_1^{(1)} + p_2' d_2 + \ldots + p_{k-1}' d_{k-1} + p_k d_k = 0$$

where now $0 \le p_i' < p_1$ for $i = 2,\ldots,k-1$. Repeat the procedure to obtain a new set of generators $d_1^*, d_2^*, \ldots, d_{k-1}^*, d_k$ which satisfy

$$pd_1^* + p_k d_k = 0.$$

If $p = 1$ then k is not lowest possible. If $p > 1$ then, cancelling as necessary, we may assume p prime to p_k. Then there are integers a,b such that $ap_k - bp = 1$ so that

$$p(ad_1^* + bd_k) + d_k = 0$$

and $\frac{1}{p}d_k \in D$. In other words the generator d_k can be replaced by the shorter generator $\frac{1}{p}d_k$. Again, repetition of the argument shows that k is not lowest possible.

Remark Lemma 3, the above proof of which I owe to J. D. Jarratt, has as corollaries several results which are often taken to be geometrically obvious. These include:

(i) The intersection of D with a s-dimensional subspace of R^n is again a discrete subgroup D' and rank D' $\leq$ s.

(ii) The definition of space group is equivalent to : G is a group of n-dimensional isometries for which the lattic T is discrete and spans R^n.

(iii) The definition of space group is equivalent to : G is a group of n-dimensional isometries for which the lattice T contains a non-zero translation vector of minimum length and spans R^n.

Theorem (Bieberbach) Let G be a discrete subgroup of $\mathcal{G}$ with $\mathcal{G}$/G compact. Then G is a space group.

Proof Since G is discrete so is its subgroup T. By Lemma 3 there is a linearly independent set $\{t_1,...,t_r\}$ which forms an integral basis for T and spans a r-dimensional subspace W of R^n. But R^n/T is a closed subspace of $\mathcal{G}$/G and therefore compact. There is an onto map $R^n/T \to R^n/W$ and so the latter quotient space is compact. This can happen only if r = n.

Remarks (i) The same conclusion would follow if the assumption on $\mathcal{G}$/G were replaced by the assumption : V/G is compact.

(ii) Bieberbach's theorem can be regarded as a topological characterisation of space groups; in the next section we obtain an analogous algebraic characterisation.

3.2. ALGEBRAIC PROPERTIES

Theorem Let G be a space group. The subgroup T is determined uniquely as the only normal free abelian subgroup, of rank n and finite index, which is maximal abelian.

<u>Proof</u> By Lemma 3 of 3.1 we know that T is normal free abelian of rank n. Suppose $T \subset S$ where S is a larger abelian subgroup. If $(u,\theta) \in S\backslash T$ then $(\theta t,\iota) = (u,\theta)(t,\iota)(u,\theta)^{-1} = (t,\iota)$ for all $t \in T$ because S is abelian. But T spans R^n and therefore $\theta = \iota$ and $(u,\theta) \in T$. Thus T is maximal abelian. To prove the uniqueness of T we suppose that S is another normal free abelian subgroup, of rank n and finite index, which is maximal abelian. If $(u,\theta) \in S$ then $(t,\iota)(u,\theta)(t,\iota)^{-1} = (t+u-\theta t,\theta)$ must still be in S (which is normal) and so commutes with (u,θ). Thus

$$u + \theta(t+u-\theta t) = (t+u-\theta t)+\theta u , \qquad t \in T$$

and $\theta^2 t-2\theta t+t = 0$ for all $t \in T$. Since T spans R^n this means $\theta = \iota + \alpha$ where $\alpha^2 = 0$. Recall that θ is orthogonal. The equation $0 = \theta'\theta-\iota = \alpha'+\alpha+\alpha'\alpha$ gives (multiplying by α)

$$\alpha'\alpha = \alpha'+\alpha = 0.$$

Then $u \in \ker\alpha \Longrightarrow \alpha u = 0$

$$u \in (\operatorname{im}\alpha)^{\perp} \Longrightarrow \underset{\text{for all } v}{u'\alpha v = 0} \Longrightarrow \underset{\text{for all } v}{u'\alpha' v = 0} \Longrightarrow \underset{\text{for all } v}{(\alpha u)'v = 0} \Longrightarrow \alpha u = 0$$

Since $\alpha^2 = 0$ the spaces $\ker\alpha$ and $(\operatorname{im}\alpha)^{\perp}$ span R^n and so $\alpha = 0$. It follows that $\theta = \iota$ and (u,θ) is a translation. Therefore (u,θ) commutes with all elements of T. But T is maximal abelian and so $(u,\theta) \in T$. Therefore S coincides with T.

<u>Corollary</u> Let G,G' be space groups. Any isomorphism $G \to G'$ restricts to an isomorphism $T \to T'$. Hence (by 2.4) any isomorphism $G \to G'$ is induced by the conjugation $\gamma \mapsto f f^{-1}$ where f is an affine transformation.

This corollary provides a proof of:

<u>Theorem</u> Let G,G' be space groups. They are isomorphic if and only if they are conjugate in A.

A similar method permits the algebraic characterisation of space groups promised in 3.1:

<u>Theorem</u> Let G_o be an (abstract) group with a normal free abelian subgroup T_o which is maximal abelian and which has finite index. Then G_o is isomorphic to a space group.

<u>Proof</u> Let n = rank T_o and $H_o = G_o/T_o$. Since T_o is normal there is an action of H_o on T_o by conjugation. It has the form $[\gamma]e_i = \sum_{j=1}^{n} a_{ij}e_j$ where $a_{ij} \in Z$ and where $\{e_1,\ldots,e_n\}$ is a basis for T_o. It is non-trivial because if $[\gamma]e_i=e_i$ for i = 1,...,n then γ commutes with all $t \in T_o$. Since T_o is maximal abelian $\gamma \in T_o$ and $[\gamma] = \iota$. Using the basis $\{e_1,\ldots,e_n\}$ we may embed $T_o \subset R^n$ and choose a scalar product so that the action of H_o is orthogonal (see Remark (iii) of 2.4). Then $H_o \subset O(n)$. It remains to prove the existence of vectors $u_i \in R^n$ for each $\theta_i \in H_o$ so that G_o may be identified with the group of isometries generated by T_o and (u_i,θ_i), i = 1,...,n. This is done in the following four steps.

First choose u_i arbitrarily. Choose also $\gamma_i \in G$ such that $[\gamma_i] = \theta_i$. Since G_o is a union of cosets $T_o\gamma_i$ the map

$$\begin{array}{ccc} G & \longrightarrow & G \\ \gamma_i & \longmapsto & (u_i,\theta_i) \end{array}$$

is well-defined. It is not however a homomorphism; its failure to be so is measured for each pair $\theta_i,\theta_j \in H_o$ by the vector $t_{ij} \in R^n$ where $\theta_i\theta_j = \theta_k$ and

$$(t_{ij}, \iota) = (u_i,\theta_i)(u_j,\theta_j)(u_k,\theta_k)^{-1},$$

$$t_{ij} = u_i + \theta_i u_j - \theta_k.$$

Secondly suppose $\theta_\ell = \theta_h\theta_i$, $\theta_k = \theta_i\theta_j$, $\theta_m = \theta_\ell\theta_j = \theta_h\theta_k = \theta_h\theta_i\theta_j$, and verify the relationship

$$t_{hi} + t_{\ell j} - t_{hk} - \theta_h t_{ij} = 0.$$

This comes from the calculation (in which for convenience we write γ_i for

$(u_i, \theta_i))$

$$(\gamma_h\gamma_i\gamma_\ell^{-1})(\gamma_\ell\gamma_j\gamma_m^{-1})(\gamma_h\gamma_k\gamma_m)^{-1}(\gamma_h\cdot\gamma_i\gamma_j\gamma_k^{-1}\cdot\gamma_h^{-1})^{-1} = \gamma_h\gamma_i\gamma_j\gamma_k^{-1}\gamma_h^{-1}\gamma_h\gamma_k\gamma_j^{-1}\gamma_i^{-1}\gamma_h^{-1}$$

$$= \iota\ .$$

Thus $\{t_{ij}\}$ defines a cocycle $t \in C^2(H_o, R^n)$ which by the divisibility theorem (see Appendix §4) is a coboundary, i.e. there exist vectors w_i such that $t_{ij} = w_i + \theta_i w_j - w_k$.

Thirdly replace the vectors u_i by $u_i - w_i$. The equations for t_{ij} now imply that the map $G \to G$ generated by $\gamma_i \in (u_i - w_i, \theta_i)$ is a homomorphism.

Finally check that this homomorphism is one-one. If $\gamma \longmapsto (0, \iota)$ then $\gamma \in T_o$ but the restriction to T_o is by construction a monomorphism. Therefore $\gamma = \iota$.

3.3. CRYSTALLOGRAPHIC RESTRICTION

Let G be a space group. By 3.1 it determines a pair (T,H) where T spans R^n and $H \subset O(n)$ is finite. Since T admits an integral basis there is an invertible λ such that $\lambda H \lambda^{-1}$ is a group of integer matrices. Conversely given a finite subgroup of GL(n,Z) we may choose a scalar product so that the action is orthogonal (see Remark (iii) of 2.4). The first consequence of this fact is the following theorem, known as the crystallographic restriction. $\phi(m)$ denotes the number of integers $0 \le d < m$ prime to m.

Theorem If H contains an element θ of order m then $n \ge \phi(m)$.

Proof The characteristic equation of θ has n roots. It is however equal to the characteristic equation of $\lambda\theta\lambda^{-1}$ which has integer coefficients and so is invariant under a permutation of primitive m-th roots of unity. Since at least one primitive m-th root satisfies the equation so must all the $\phi(m)$ primitive m-th roots. Therefore $n \ge \phi(m)$.

Remarks (i) If $m > 2$ then $\phi(m)$ is even and the restriction on possible m is the same for odd $n = 2k+1$ as for even $n = 2k$. For example $n = 2,3$ implies $m = 1,2,3,4$ or 6. Similarly $n = 4,5$ implies $m = 1,2,3,4,5,6,8,10$ or 12.

(ii) The restriction is satisfied automatically by odd $m \le n+1$ and by even

$m \leq 2n$. On the other hand an odd prime m in the range $n+1 < m$ cannot satisfy the restriction because then $\phi(m) = m-1$. The restriction is not a sufficient condition for H to act on a lattice of rank n, for such actions impose restrictions on the number of generators of H as well as on their orders. It is however the starting point for the proof of the Jordan-Zassenhaus Theorem: the number of conjugacy classes of finite subgroups $H \subset GL(n,Z)$ for given n is finite [60].

A closely related result is due to Minkowski [34]. He computed the least common multiple of the orders of all finite subgroups $H \subset GL(n,Z)$. His starting point is the fact that the scalar product defines a quadratic form f invariant under H together with the following

<u>Lemma</u> Let $\theta \in GL(n,Z)$ be of finite order and suppose $\theta \equiv \iota$ modulo p where p is an odd prime. Then $\theta = \iota$. [For the proof see §1 of [34]].

The lemma implies that the subgroup H is mapped monomorphically to a subgroup of $GL(n,Z/pZ)$, also leaving the quadratic form f invariant. The collection of subgroups of $GL(n,Z/pZ)$ leaving some f invariant has orders dividing

$$\text{ה} = \prod_{\substack{q \\ \text{prime}}} q^{[\frac{n}{q-1}]+[\frac{n}{q(q-1)}]+[\frac{n}{q^2(q-1)}]+\cdots}$$

Moreover for any given prime q it is possible to find a quadratic form f for which the order of the group H leaving f invariant has the same power of q as ה . Therefore ה is the least common factor of the orders of finite subgroups $H \subset GL(n,Z)$ and hence of the orders of finite H which can act on a lattice T of rank n. Note that

$$n! = \prod_{\substack{q \\ \text{prime}}} q^{[\frac{n}{q}]+[\frac{n}{q^2}]+[\frac{n}{q^3}]+\cdots}$$

and therefore ה is divisible by $n!2^n$ with odd quotient. On the other hand ה divides $(2n)!$ (consider the effect of replacing each $q-1$ by $\frac{1}{2}q$ in the expression for ה).

3.4. ARITHMETIC CRYSTAL CLASSES

It is natural to arrange the equivalence classes of space groups according to some coarser classification. This can be done in various ways, all of which involve the action of H on T, which is denoted (T,H) in the sequel.

Definition The actions (T,H),(T',H') are *arithmetically equivalent* if there is an invertible linear transformation λ such that $T'=\lambda T$, $H'=\lambda H\lambda^{-1}$ and the action of H' on T' is induced by that of H on T. An equivalence class is called an arithmetic crystal class.

By 3.3 there is a one-one correspondence between arithmetic crystal classes and conjugacy classes of finite subgroups of GL(n,Z), and the number of arithmetic classes for given dimension n is finite.

The results of 3.2. show that if G,G' are isomorphic then there is an affine transformation $f = (w,\lambda)$ such that $G' = fGf^{-1}$ and hence $T' = T, \lambda H' = \lambda H\lambda^{-1}$. Thus (T,H),(T',H') are arithmetically equivalent. Therefore there is a map

$$\text{isomorphism classes of space groups} \longrightarrow \text{arithmetic crystal classes}$$

The map is always onto because, given a pair (T,H) where T spans R^n and $H \subset O(n)$ is finite, there is a space group G defined by $G = \{(t,\theta);\ t \in T,\ \theta \in H\}$. Some indication of the number of space groups attached to each arithmetic crystal class is given by the following statistics

n	1	2	3	4	5
number of arithmetic crystal classes	2	13	73	710	?
number of isomorphism classes of space groups	2	17	219	4783	?
" in largest arithmetic crystal class	1	3	16	272	11456
" in next largest arithmetic crystal class	1	2	16	218	9608
" in smallest arithmetic crystal class	1	1	1	1	1

It will be seen that typically there are a few very large arithmetic crystal classes together with a rather large number of small arithmetic crystal classes. This is the reason for concentrating attention on certain large crystal classes in Chapter Six.

<u>Theorem</u> The number of isomorphism classes of space groups attached to a given arithmetic crystal class (T,H) is equal to the number of orbits of the action of N(H) - the normaliser of H in GL(n,R) - on the cohomology group $H^2(H,T)$.

<u>Proof</u> To construct a space group G it is necessary and sufficient to specify vectors $v_i \in R^n$ such that $(v_i,\theta_i) \in G$ for $\theta_i \in H$. Note that if v_i is replaced by t_i+v_i for some $t_i \in T$ then G is unchanged; similarly if $w \in R^n$ is a fixed vector then if v_i is replaced by $v_i+w-\theta_i w$ then G is replaced by the equivalent group $(w,\iota)G(w,\iota)^{-1}$; the choice of v_i is not arbitrary since $\theta_i\theta_j = \theta_k$ implies $v_k \equiv v_i+\theta_i v_j$ mod T. Putting these facts together we see that $\{v_i\}$ determines an element $v \in H^1(H,R^n/T)$. The divisibility theorem shows $H^1(H,R^n) = H^2(H,R^n) = 0$ and therefore the cohomology exact sequence yields an isomorphism $H^1(H,R^n/T) \cong H^2(H,T)$; in fact v is mapped to the element t defined by $t_{ij} = v_i+\theta_i v_j-v_k$ encountered already in 3.2. Conversely a cohomology element $t \in H^2(H,T)$ or equivalently $v \in H'(H,R^n/T)$ will determine a unique isomorphism class of space groups. It remains to answer the question: when could distinct cohomology elements $v,v' \in H^1(H,R^n/T)$ determine isomorphic space groups? Only when there is an invertible linear transformation λ such that

$$(w,\lambda)(v_i,\theta_i)(w,\lambda)^{-1} = (v_i',\theta_i') ,$$

that is

$$\theta_i' = \lambda\theta_i\lambda^{-1} \text{ and } v_i' = w+\lambda v_i - \lambda\theta_i\lambda^{-1}w \sim \lambda v_i,$$

where ~ means *cohomologous*. In other words $\lambda \in N(H)$ and $v' = \lambda\cdot v$ under the induced action

$$\begin{array}{ccc} H^1(H,R^n/T) & \xrightarrow{\lambda} & H^1(H,R^n/\lambda T) \\ \{v_i\} & \longmapsto & \{\lambda v_i\} \end{array}$$

<u>Corollary</u> The number of isomorphism classes of space groups for given dimension n is finite.

<u>Proof</u> By the finiteness of $H^2(H,T)$ and of the number of arithmetic crystal classes for given n.

3.5. BRAVAIS TYPES, GEOMETRICAL CRYSTAL CLASSES AND CRYSTAL FAMILIES

The number of arithmetic crystal classes is large, even for n=3, and in practice crystallographers have used a still coarser definition. There are choices involved in deciding how to extend such methods to arbitrary dimensions, and hence some disagreement as to the correct definitions to adopt (it is no longer possible to use mineralogical evidence as a guide!). We shall adopt the following which seem reasonably natural.

<u>Definition</u> The lattices T,T' are *arithmetically equivalent* if the actions (T,G(T)), (T',G(T')) are arithmetically equivalent. Here $G(T) = \{\theta \in O(n);\ \theta T = T\}$. An equivalence class is called a *Bravais type*.

<u>Definition</u> The actions (T,H),(T',H') are *geometrically equivalent* if there is an invertible linear transformation λ such that $H' = \lambda H \lambda^{-1}$. An equivalence class is called a *geometric crystal class*. The lattices T,T' are geometrically equivalent if the actions (T,G(T)), (T',G(T')) are so, i.e. $G(T') = \lambda G(T) \lambda^{-1}$. An equivalence class is called a *Bravais system*.

The above definitions imply the existence of maps (denoted in the diagram by solid arrows)

$$\begin{array}{ccc}
\{\text{arithmetic crystal classes}\} & \longrightarrow & \{\text{geometric crystal classes}\} \\
\cup\uparrow\ \vdots\downarrow & & \cup\uparrow \\
\{\text{Bravais types}\} & \longrightarrow & \{\text{Bravais systems}\}
\end{array}$$

Here all the sets are finite and vertical arrows denote natural inclusions. In practice, crystallographers wish to assign each arithmetic crystal class to a definite Bravais type and Bravais system. This means defining a map corresponding to the dotted arrow.

Remark The proposal $(T,H) \longmapsto (T,G(T))$ for such a map is inadequate because it is not well defined. Among the pairs (T,H) in a given arithmetic crystal class there will be some for which, because of special values of the lattice parameters, T has especially high symmetry; such symmetry is unstable under variation of the lattice parameters and is not physically significant. We therefore introduce a partial ordering on the lattices T which occur in a given arithmetic crystal class.

Definition Suppose that (T,H), (T',H') are arithmetically equivalent. Let $T \geq T'$ if there is an invertible linear transformation λ such that $\lambda T = T'$, $\lambda H \lambda^{-1} = H'$ and $\lambda G(T) \lambda^{-1} \supseteq G(T')$.

Note that T and T' have the same Bravais type if and only if $T \geq T'$ and $T' \geq T$. The desired map will be achieved if it can be shown that each arithmetic crystal class contains a pair (T_o,H_o) for which T_o is *minimal* i.e. $T_o \leq T$ for all T occurring in pairs (T,H) of the arithmetic crystal class. Since the number of Bravais types is finite it is sufficient to prove that if (T_1,H_1), (T_2,H_2) are arithmetically equivalent then there exists (T_3,H_3) in the same class with $T_3 \leq T_1$, $T_3 \leq T_2$.

Lemma Let H_1,H_2 be the subgroups of $O(n)$ such that $\psi H_1 \psi^{-1} = H_2$ for some $\psi \in GL(n,R)$. Then $\psi = \theta\alpha$ where $\theta \in O(n)$ and α is positive definite, $\theta H_1 \theta^{-1} = H_2$ and α commutes with every element of H_1. If $k > 0$ is an integer then $(\alpha^{1/k}T_1,H_1)$ is arithmetically equivalent to (T_1,H_1).

Proof The first statements follow from the argument used already in 2.3. Since α commutes with $\theta_1 \in H_1$ so does $\alpha^{1/k}$. Therefore $H_1 = \alpha^{1/k}H_1\alpha^{-1/k}$ and H_1 acts on $\alpha^{1/k}T_1$.

Theorem Let (T_1,H_1), (T_2,H_2) be arithmetically equivalent. There is an action (T_3,H_3) in the same arithmetic crystal class with $T_3 \leq T_1$, $T_3 \leq T_2$.

Proof Let $\psi = \theta\alpha$ be as in the lemma and $T_3 = \alpha^{1/k}T_1$ where k is chosen sufficiently large (i.e. with $\alpha^{1/k}$ sufficiently close to the identity) to make $G(T_3) = G(\alpha^{1/k}T_1)$ equal to $\alpha^{1/k}G(T_1)\alpha^{-1/k} \cap O(n)$. Then $T_3 \leq T_1$. In fact $G(T_3) = \alpha^{1/k} H_1' \alpha^{-1/k}$ for some subgroup $H_1' \subset G(T_1)$. By the lemma $\alpha^{1/k}$ commutes with each element of H_1' and hence $G(T_3) = H_1'$. Now choose $H_3 = \alpha^{1/k} H_1 \alpha^{-1/k}$. Then if $\mu = \psi\alpha^{-1/k}$ we have $T_2 = \mu T_3$ and $H_2 = \mu H_3 \mu^{-1}$ while

$$\mu G(T_3)\mu^{-1} = \psi H_1' \psi^{-1} = \theta H_1' \theta^{-1}$$

lies in $O(n)$ and is therefore a subgroup of $G(T_2)$. Therefore $T_3 \leq T_2$.

Warning There is no corresponding construction of a map associating each geometric crystal class to a Bravais system and the attempt in [48] to produce such a map is fallacious. If it is desired to have such a map which is consistent with that assigning arithmetic crystal classes to Bravais types then the definition of crystal system must be modified. This is shown by examples when n=3 (see [48] and Chapter 5).

The maps, and corresponding equivalence relations R_1, R_2 on arithmetic classes, make it natural to define the *push-out* set, that is the quotient set under the weakest equivalence relation which includes both R_1 and R_2. This is the set of families. In view of the existence of minimal lattices

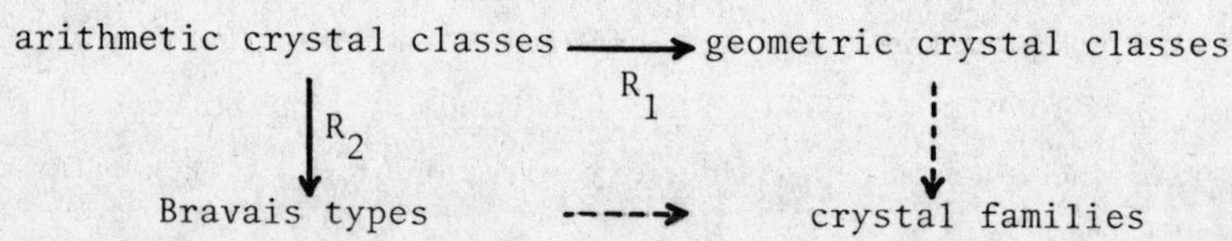

for each arithmetic crystal class the notion can be defined as follows.

Definition A *crystal family* is an equivalence class of lattices which span R^n under the relation: $T_1 \sim T_2$ if there exist pairs (T_1, H_1) and (T_2, H_2) with T_i a minimal lattice for the arithmetic class of (T_i, H_i) for i=1,2 and with $H_2 = \psi H_1 \psi^{-1}$ for some invertible linear transformation.

3.6. COLOUR SYMMETRY GROUPS

Consider a space group G together with a representation of G as a transitive group of permutations of N objects, that is, a homomorphism to the symmetric group

$\rho : G \to S_N$.

Such a structure occurs whenever G is viewed as a group of symmetries of an object painted in N colours, the elements of G being isometries which permute the colours in a consistent manner. If $G_i \subset G$ is the subgroup of all elements of G which leave unchanged the i-th colour then transitivity implies

$$G_j = \gamma G_i \gamma^{-1}$$

for some $\gamma \in G$ chosen so that $\rho(\gamma)$ maps i to j. Conversely given the single subgroup $G_1 \subset G$ the representation ρ may be reconstructed: denote by $1,\ldots,N$ the coset of G_1 in G and associate to $\alpha \in G$ the permutation $\rho(\alpha)$ of $\{1,\ldots,N\}$ which sends the coset βG_1 to the coset $\alpha\beta G_1$. It is therefore convenient to define colour symmetry by means of the subgroup G_1 rather than by means of the representation ρ.

Definition A *colour symmetry group* is a pair consisting of a space group G and a subgroup G_1 of finite index. The index N is called the *number of colours*. Two colour symmetry groups $G_1 \subseteq G$ and $G_1' \subseteq G'$ are *equivalent* if there is an affine transformation f such that $G_1' = fG_1f^{-1}$ and $G' = fGf^{-1}$.

A number of consequences of the definition are almost immediate from the results of this chapter.

Proposition G_1 is a space group.

Proof G_1 is discrete and its lattice T_1, being of finite index in T, must span R^n.

Proposition If there is an isomorphism $G \to G'$ which maps $G_1 \to G_1'$ then the colour symmetry groups $G_1 \subset G$ and $G_1' \subset G'$ are equivalent.

Proof By 3.2 there is an affine transformation f such that the isomorphism $G \to G'$ is induced by conjugation by f. It follows that $G_1' = fG_1f^{-1}$.

Proposition $T_1 = T \cap G_1$ and there is a diagram of exact sequences

$$\begin{array}{ccccccccc} 0 & \longrightarrow & T & \longrightarrow & G & \xrightarrow{r} & H & \longrightarrow & 1 \\ & & \cup & & \cup & & \cup & & \\ 0 & \longrightarrow & T_1 & \longrightarrow & G_1 & \longrightarrow & H_1 & \longrightarrow & 1 \end{array}$$

where $\cup$ denotes inclusion.

Proof Immediate from the definitions.

Theorem Let $G_1 \subseteq G$ be a colour symmetry group. There is a unique space group G' such that $G_1 \subseteq G' \subseteq G$ and
(i) $G_1 \subseteq G'$ is a colour symmetry group with $T_1 \subseteq T'$ and $H_1 = H'$,
(ii) $G' \subseteq G$ is a colour symmetry group with $T' = T$ and $H' \subseteq H$.

Proof Let $G' = r^{-1}H_1$. Then $G_1 \subseteq G' \subseteq G$ has the required properties. Conversely if $G_1 \subseteq G' \subseteq G$ satisfies (i), (ii) then $G' = r^{-1}H_1$.

The cases (i) and (ii) are called the *klassengleich* and *zellengleich* cases respectively. The names are due to C. Hermann and could be translated as *class equivalent* and *lattice equivalent*. If $T_1 \subseteq T'$ has index r and $H' \subseteq H$ has index s then $G_1 \subseteq G$ has index $N = rs$. In this way any colour symmetry group is built up from two elementary colour symmetry groups corresponding to the cases (i) and (ii). The two cases have very different characteristics and are best treated separately, as in the following definitions.

Definition A *class equivalent colour symmetry group* is a pair of space groups $G_1 \subseteq G$ with $T_1 \subseteq T$ of finite index and $H_1 = H$.

It follows from this definition that T_1 is a H-submodule of T, and that the elements $t_1 \in H^2(H,T_1)$ and $t \in H^2(H,T)$ which define G_1 and G (see 3.4) are consistent, i.e. t_1 is mapped to t in the exact sequence induced by the inclusion $T_1 \subseteq T$.

$$H^1(H,T/T_1) \to H^2(H,T_1) \to H^2(H,T) \to H^2(H,T/T_1)$$
$$t_1 \longmapsto t$$

We obtain

<u>Proposition</u> Let r be the index of T_1 in T. For given r the number of class equivalent colour symmetry groups is finite. For r prime to $|H|$, there is one for each space group G and choice of H-invariant sublattice $T_1 \subseteq T$ of index r.

<u>Proof</u> The first statement follows from the finiteness of $H^2(H,T)$ and $H^1(H,T/T_1)$; the second from the fact that if r is prime to $|H|$ then T/T_1 is divisible by $|H|$ and hence $H^1(H,T/T_1) = H^2(H,T/T_1) = 0$ showing that t_1 is determined uniquely by t.

<u>Warning</u> It may be that there are no H-invariant sublattices $T_1 \subset T$ of rank r and hence no class equivalent colour symmetry groups of this index. For examples see Chapter 5.

<u>Definition</u> A *lattice equivalent colour symmetry group* is a pair of space groups $G_1 \subseteq G$ with $T_1 = T$ and $H_1 \subseteq H$.

It follows from this definition that $H_1 \subseteq H$ has finite index s dividing the order of H, and that $t_1 \in H^2(H_1,T)$ is determined uniquely by $t \in H^2(H,T)$ as the image under the restriction homomorphism

$$H^2(H,T) \to H^2(H_1,T)$$
$$t \longmapsto t_1$$

We obtain

<u>Proposition</u> The number of lattice equivalent colour symmetry groups is finite. There is one for each space group G and choice of conjugacy class of subgroups $H_1 \subseteq H$.

Putting together the results of the above propositions we have

Theorem The number of isomorphism classes of colour symmetry groups $G_1 \subseteq G$, with given index N and given dimension n, is finite.

Remark When N=2 we shall use the phrase *black and white group* to mean a colour symmetry group $G_1 \subset G$ or point group $H_1 \subset H$ as appropriate.

4 Point groups

It was shown in the previous chapter that to study space groups it is necessary to study arithmetic crystal classes or actions (T,H). This can be done in two ways. Either study all finite subgroups $H \subset O(n)$ and ask which can act on a lattice T; this is the approach adopted in the present chapter which follows that of [58]. Or study all lattices $T \subset R^n$ and ask for subgroups H of the symmetry group G(T); this is done in the next chapter.

4.1. FINITE SUBGROUPS OF O(n)

Let O(n) denote the group of orthogonal transformations in dimension n, and SO(n) the subgroup of those with determinant 1. Since $\theta \in O(n)$ has $1 = \det(\theta'\theta) = (\det\theta)^2$ we have

$$O(n) = SO(n) \cup \beta\, SO(n)$$

for some choice of β with $\det\beta = -1$. Similarly if $H \subset O(n)$ is a finite subgroup there are two possibilities: either $H \subset SO(n)$ or $H = H_o \cup \beta H_o$ where $H_o \subset SO(n)$, $\det\beta = -1$, and $H_o = \beta H_o \beta^{-1}$.

In the special case when n is odd more can be said because then $O(n) = SO(n) \cup \bar{\iota}SO(n)$ where $\bar{\iota}$ is multiplication by -1 (we shall follow the example of crystallographers in writing *minus* above rather than before a number or symbol, and in denoting the group generated by $\bar{\iota}$ by $\bar{1}$; since complex conjugation never occurs in these notes there is no risk of confusion). Now if $\det\beta = -1$ then $\beta = \bar{\iota}\cdot\beta$ where $\beta\bar{\iota} \in SO(n)$. Therefore $H = H_o \cup \beta H_o$ is either $H_o \cup \bar{\iota}H_o = H_o \times \bar{1}$ or else $H_o \cup \bar{\iota}(H_1 \backslash H_o)$ where $H_o \subset H_1 \subset SO(n)$ is a *black and white group* i.e. group H_1 with subgroup H_o of index 2. Thus when n is odd the enumeration of finite subgroups of O(n) is determined by that of subgroups (including black and white groups) of SO(n).

When n=1 there are only two subgroups of O(1) because SO(1) is trivial. Explicitly these are

$1 = \{\iota\}$

$\bar{1} = \{\iota, \bar{\iota}\}$

When n=2 any $\theta \in O(2)$ has eigenvalues $e^{\pm i\alpha}$, corresponding to rotation through α, or ± 1 corresponding to reflection in a line.

<u>Lemma</u> Let H_o be a finite subgroup of SO(2). Then $H_o = \{\iota, \theta, \ldots, \theta^{q-1}\}$ where θ is rotation through $2\pi/q$.

<u>Proof</u> If H_o contains rotations through $2\pi/q_1$ and $2\pi/q_2$ then it contains a rotation through $2\pi/q$ where q is the lowest common multiple of q_1 and q_2.

<u>Theorem</u> Any finite subgroup of O(2) has the form

$$q = \{\iota, \theta, \ldots, \theta^{q-1};\ \theta^q = \iota\}$$

or

$$qm = \left\{ \begin{matrix} \iota, \theta, \ldots, \theta^{q-1} \\ \beta, \beta\theta, \ldots, \beta\theta^{q-1} \end{matrix} ;\ \theta^q = \beta^2 = (\beta\theta)^2 = \iota \right\}$$

where θ is rotation through $2\pi/q$ and β is reflection in a line.

<u>Proof</u> The result follows immediately from the lemma. Note that, although qm depends on the choice of β, any two choices are conjugate.

<u>Remarks</u> (i) The notation is that introduced into crystallography by Hermann-Mauguin and is a shorthand listing of convenient generators. It is not unique, since 2 could be written as $\bar{1}$ (thinking of θ as $\bar{\iota}$ instead of rotation through π) and 6 could be written as $\bar{3}$ (thinking of rotation through $2\pi/6$ as the product of rotation through $2\pi/3$ and $\bar{\iota}$). Note that q is an integer whereas m denotes *mirror*.

(ii) Only 10 of the conjugacy classes satisfy the crystallographic restriction (see 3.3), namely

1,2,3,4,6,1m,2m,3m,4m,6m.

It is usual to write m for 1m, and mm for 2m indicating a choice of two reflections in perpendicular lines as generators. Sometimes authors write 4mm for 4m, thus drawing attention to the reflection $\beta\theta$ in a line at angle $\frac{\pi}{4}$ to the mirror line of β; similarly 6mm for 6m. Further discussion of the Hermann-Mauguin notation is given in Chapter Six.

4.2. FINITE SUBGROUPS OF SO(3)

The eigenvalues of $\theta \in SO(3)$ must have product 1 and therefore take the form 1, $e^{\pm i\alpha}$. Thus θ has a single real eigenvector u, called its *pole*, and is a rotation about u through an angle $2\pi/q$ if θ has order q. Clearly if u is a pole for some elements of a finite subgroup $H \subset SO(3)$ then so is $\bar{u}$. It follows that if $|H| = N$ and if q(u) is the order of u then

$$N - 1 = \sum_{u} \tfrac{1}{2}(2(u)-1)$$

the sum being taken over all poles of H.

On the other hand if $\psi \in H$ and u is a pole of order q then ψu is again a pole of order q. In fact if u is a pole for θ then u is a pole for $\psi\theta\psi^{-1}$. The number of distinct poles obtained in this way is equal to the number of cosets N/q(u) of the subgroup $\{\iota,\theta,\ldots.,\theta^{q-1}\}$. Therefore

$$N - 1 = \sum_{[u]} \frac{N}{q(u)} \tfrac{1}{2}(q(u)-1)$$

where now the summation is over equivalence classes of poles formed by the relation $u \sim u'$ if $u' = \psi u$ for some $\psi \in H$. If N = 1 then $H = \{\iota\}$ is the subgroup denoted 1. If $N \geq 2$ the relation becomes

$$2(1 - \frac{1}{N}) = \sum_{[u]} (1 - \frac{1}{q(u)}) \qquad (*)$$

and there must be either two or three terms in the summation, since only one term would imply $2(1 - \frac{1}{N}) < 1$ and more than three terms would imply $2(1 - \frac{1}{N}) \geq 2$. If there are three terms then at least one has q(u) = 2 since otherwise $2(1 - \frac{1}{N}) \geq 2$.

Proposition If the number of terms in the summation (*) is two then H has the form

$$q = \{\iota, \theta, \ldots, \theta^{q-1};\ \theta^q = \iota\}$$

where θ is a rotation through $2\pi/q$ about some axis.

Proof We have $2 = \frac{N}{q(u_1)} + \frac{N}{q(u_2)}$, an equation which can be satisfied only if $q(u_1) = q(u_2) = N$. Therefore the two poles must be $\pm u$ for some u, and H must consist of rotations about this axis. Any other choice of axis will give a conjugate group.

Proposition If the number of terms in the summation (*) is three, and if $q(u) = 2$ for at least two terms, then H has the form

$$q22 = \left\{ \begin{matrix} \iota, \theta, \ldots, \theta^{q-1} \\ \rho, \rho\theta, \ldots, \rho\theta^{q-1} \end{matrix} ;\ \theta^q = \rho^2 = (\rho\theta)^2 = \iota \right\}$$

where θ is rotation through $2\pi/q$ about an axis $\pm u$ and $\rho, \rho\theta$ are rotations through π about axes (angle $\frac{\pi}{q}$ apart) perpendicular to u.

Proof Let q be the order of poles in the third term. Each pole in this equivalence class must be equivalent to its negative. But $2(1 - \frac{1}{N}) = 1 - \frac{1}{q} + 1$ and so $N = 2q$. Therefore the equivalence class of order q contains exactly two poles $\pm u$. The other families each contain q poles of order 2.

Proposition If the number of terms in the summation (*) is three, and if $q(u) = 2$ for only one term then H has one of the forms

$$233 = \{\phi, \psi;\ \phi^2 = \psi^3 = (\phi\psi)^3 = \iota\} = \text{ rotations of tetrahedron,}$$
$$432 = \{\phi, \psi;\ \phi^4 = \psi^3 = (\phi\psi)^2 = \iota\} = \text{ rotations of cube,}$$
$$532 = \{\phi, \psi;\ \phi^5 = \psi^3 = (\phi\psi)^2 = \iota\} = \text{ rotations of icosahedron.}$$

Proof If the three terms correspond to poles of order q,r,2 then

$\frac{2}{N} = \frac{1}{q} + \frac{1}{r} - \frac{1}{2}$. At least one of q,r must be 3 since otherwise $q,r \geq 4$ and $\frac{2}{N} \leq 0$, and the other must be 3,4 or 5 giving N = 12,24 or 60 respectively. We consider these cases in turn.

N = 12 Four poles of order 3, say u_1,u_2,u_3,u_4 form an equivalence class. Their negatives $\bar{u}_1,\bar{u}_2,\bar{u}_3,\bar{u}_4$ form a different equivalence class. The action of H permutes u_1,u_2,u_3,u_4 which therefore form a regular tetrahedron. Since every $\theta \in H$ has $\det \theta = 1$ the induced permutation is even giving an isomorphism $H \to A_4$. Generators are $\phi = (12)(34)$, a rotation through π about $\frac{1}{2}(u_1+u_2) = -\frac{1}{2}(u_3+u_4)$, $\psi = (234)$ and $\phi\psi = (412)$.

N = 24 Eight poles of order 3, say $\pm u_1, \pm u_2, \pm u_3, \pm u_4$ form an equivalence class. The action of H permutes these four axes which therefore form the diagonals of a cube, giving an isomorphism $H \to S_4$. Generators are $\phi = (1234), \psi = (214)$ and $\phi\psi = (34)$.

N = 60 Twelve poles of order 5, say $\pm u_1, \pm u_2, \pm u_3, \pm u_4, \pm u_5$ which must form the vertices of an icosahedron, and twenty poles of order 3 which form the vertices of a dodecahedron. If $\phi^5 = \psi^3 = \iota$ then $\phi\psi$ can be calculated by noting that the 20 poles can be arranged as 5 tetrahedra and that any action of the group gives an even permutation of these tetrahedra, giving an isomorphism $H \to A_5$. Generators are $\phi = (12345)$, $\psi = (254)$, $\phi\psi = (12)(34)$.

We may summarise by listing all finite subgroups of SO(3) as

q	isomorphic to C_q
q2 = q22	isomorphic to D_q
23 = 233	isomorphic to A_4
43 = 432	isomorphic to S_4
53 = 532	isomorphic to A_5

Inspection shows that the possible black and white groups $H_0 \subset H_1$ in SO(3) are, writing r = 2q,

$q \subset r$

$q \subset q22$

$q22 \subset r22$

$233 \subset 432$

4.3. FINITE SUBGROUPS OF O(3) WHICH ACT ON A LATTICE

By 4.1 we must consider subgroups $H \subset O(3)$ which satisfy the crystallographic restriction, and can do so by considering the three possibilities $H = H_o$, $H = H_o \times \bar{1}$ and $H = H_o \cup \bar{\iota}(H_1\ H_o)$ where $H_o \subset SO(3)$ and $H_o \subset H_1 \subset SO(3)$.

$\underline{H = H_o}$ There are 11 possibilities

1	2	3	4	6	233
	222	322	422	622	432

$\underline{H = H_o \times \bar{1}}$ If θ has odd order q then $\bar{\theta}$ generates both $\bar{\iota}$ and θ, but if θ has even order two generators are necessary, for example θ and reflection in the plane perpendicular to θ denoted by the symbol /m. We obtain

$\bar{1}$	2/m	$\bar{3}$	4/m	6/m	m3
	mmm	$\bar{3}$m	4/m mm	6/m mm	m3m

Some of these symbols are convenient abbreviations. More correct, but pedantic, would be $\frac{2}{m}\frac{2}{m}\frac{2}{m}$, $\bar{3}\frac{2}{m}$, $\frac{4}{m}\frac{2}{m}\frac{2}{m}$, $\frac{6}{m}\frac{2}{m}\frac{2}{m}$, $\frac{4}{m}\bar{3}\frac{2}{m}$ in the second row.

$\underline{H = H_o \cup \bar{\iota}(H_1 - H_o)}$ If $\theta \in H_1 \backslash H_o$ and $\theta^2 \in H_o$ then $\bar{\theta}$ is a convenient generator of H. If however θ has order 2 it is even more convenient to consider $\bar{\theta}$ as a reflection. This explains the notation m in the following listing of 10 further groups arising from black and white groups $H_o \subset H_1$ in SO(3).

H_o	1	2	3	2	3	4	6	222	322	233
H_1	2	4	6	222	32	422	622	422	622	432
H	m	$\bar{4}$	$\bar{6}$	2mm	3m	4mm	6mm	$\bar{4}$2m	$\bar{6}$2m	$\bar{4}$3m

In all there are 32 conjugacy classes of subgroups $H \subset O(3)$ which can act on a lattice. This result of J. F. C. Hessel includes the crystallographic restriction for n=3 and was first published 150 years ago [25]. When n=3 there is a very natural assignment of these 32 geometric crystal classes to the 7 crystal systems which is summarised in the table (the underlined group being the symmetry group of a lattice) and explained further in the next chapter.

Tri- clinic	Mono- clinic	Ortho- rhombic	Tri- gonal	Tetra- gonal	Hexa- gonal	Cubic
1	2		3	4	6	23
$\bar{1}$	m		$\bar{3}$	$\bar{4}$	$\bar{6}$	
	2/m			4/m	6/m	m3
		2mm	3m	4mm	6mm	
		222	32	422	622	432
			$\bar{3}$m	$\bar{4}$2m	$\bar{6}$2m	$\bar{4}$3m
		mmm	*	4/m mm	6/m mm *	m3m

* These geometric crystal classes belong to the same family (see 5.4)

Note however that (i) the assignment is not compatible with the map from arithmetic crystal classes to Bravais types (see 3.5), and (ii) results of Neubüser and Plesken for n=7 show that no such well defined assignment is possible in general without some modification of the definition of system (see [36]).

4.4. n-DIMENSIONAL POINT GROUPS

The table displaying the 32 geometrical crystal classes suggests a way of building up point groups $H \subset O(n)$ for arbitrary dimension n. The action of H on R^n determines a decomposition of R^n as a sum of invariant subspaces and hence a partition of n. The allocation of geometric classes to partitions for n=1,2,3 is easily determined, and is indicated in the table. $\pi(n)$ denotes the number of partitions of n.

n	$\pi(n)$	partition of n	number of families	number of systems	point groups
1	1	1 = 1	(1)	(1)	$1,\bar{1} = 2$
2	2	2 = 1+1	(2)	(2)	1,2,m,2m
		2 = 2	(2)	(2)	3,4,6,3m,4m,6m
3	3	3 = 1+1	(3)	(3)	$1,\bar{1}$,2,m,2/m 2mm,222,mmm
		3 = 2+1	(2)	(3)	$3,\bar{3},3m,32,\bar{3}m$ $4,\bar{4},4/m,4mm,422,\bar{4}2m,4/mmm$ $6,\bar{6},6/m,6mm,622,\bar{6}2m,6/m\ mm$
		3 = 3	(1)	(1)	$23,m3,432,\bar{4}3m,m3m$

<u>Theorem</u> Every crystal family, and hence also every crystal system, determines a partition of n.

<u>Proof</u> Let (T,H) belong to an arithmetic class for which T is minimal in the sense of 3.5. The subgroup $H \subset O(n)$ determines a decomposition of R^n into invariant subspaces, say $R^n = V_1 \oplus \ldots \oplus V_s$ and a partition $n = n_1 + \ldots + n_s$ where $n_i = \dim V_i$. Since $H \subseteq G(T)$ every V_i is G(T)-irreducible. However if G(T) contained a symmetry mapping V_i to V_j this could be destroyed by altering T by appropriate scalar factors c_i, c_j on V_i, V_j and so T is not minimal. Therefore G(T) determines the same decomposition of R^n, and in particular the partition is invariant under the equivalence relation defining crystal families.

The relationship between crystal families and partitions goes further. Consider a particular partition of n into s components determined by H. If the structure of H implies that the summands $V_1, \ldots, V_s$ are in special position - for example, that two are perpendicular - then the same must hold for G(T) and hence for the crystal family.

<u>Example</u> Consider the partition $n = 1+....+1$. It corresponds to a decomposition $R^n = V_1 \oplus .... \oplus V_n$ where $\dim V_i = 1$. Consider the equivalence relation between components generated by $i \sim j$ if V_i perpendicular to V_j whenever H acts; this gives a subsidiary partition of n. So the number of crystal families corresponding to $n = 1+....+1$ is itself at least $\pi(n)$. Similarly the number corresponding to $n = 2+1+....+1$ is at least $\pi(n-2)$.

<u>Problem</u> Elucidate further the relationship between the number f(n) of crystal families in dimension n and the number $\pi(n)$ of partitions of n. For example, what is the behaviour of the ratio $f(n+1)/f(n)$ as $n \to \infty$, Some steps in this direction have been taken by Jarratt [27].

The enumeration of crystal families has two aspects, as was already pointed out by C. Hermann [24]. The first is to describe how families in lower dimensions combine to give families in higher dimensions. The second is to describe those families which are genuinely new in a given dimension n i.e. those with partition n=n. They arise whenever $H \subset O(n)$ acts transitively on a set which spans R^n, and has no invariant subspace.

<u>Definition</u> $H \subset O(n)$ is *fully transitive* if it induces all possible permutations on a set of k vectors which span R^n and has no invariant subspace.

<u>Examples</u> (i) Let $e_1,....,e_n$ be mutually perpendicular and of equal length. The lattice T of all integral combinations of $e_1,...,e_n$ is called *primitive hypercubic*. The symmetry group G(T) has order $2^n\, n!$ being generated by permutation maps $e_j \longmapsto e_{\sigma(j)}$ for $\sigma \in S_n$ and reflections $\mu_i : e_j \longmapsto \begin{cases} e_j, & j \neq i \\ -e_j, & j = i \end{cases}$ for $i = 1,...,n$. The group G(T), and several of its subgroups, is fully transitive.

(ii) Let $f_1,...,f_{n+1}$ be vectors of equal length which satisfy $f_1+....+f_{n+1} = 0$ such that the angle between any pair f_i, f_j is the same. These conditions imply that the angle α is determined by the equation $\sum_{i,j} (f_i, f_j) = 0$:

$$0 = \sum_i (f_i, f_i) + \sum_{i \neq j} (f_i, f_j)$$

$$= (n+1)(f_1, f_1) + n(n+1)(f_1, f_2)$$

and therefore $\cos \alpha = \dfrac{(f_1, f_2)}{(f_1, f_1)} = -\dfrac{1}{n}$. The lattice T of all integral combinations of $f_1, \ldots, f_n$ is called *primitive hyperbolic*; its symmetry group G(T) has order $2 \cdot (n+1)!$ being generated by permutation maps $f_j \longmapsto f_{\sigma(j)}$ for $\sigma \in S_{n+1}$ and by the inversion $f_j \longmapsto \bar{f}_j$. The group G(T), and several of its subgroups, is fully transitive.

(iii) It is confusing that for n=3 the above two symmetry groups G(T) are isomorphic and even determine the same family. The lattices T are distinct. The primitive hypercubic lattice is generated by (100)(010)(001) lying at the vertices of stacked cubes, while the primitive hyperbolic lattice is generated by $(\frac{1}{2}\frac{1}{2}\frac{1}{2})(\bar{\frac{1}{2}}\bar{\frac{1}{2}}\frac{1}{2})(\bar{\frac{1}{2}}\frac{1}{2}\bar{\frac{1}{2}})(\frac{1}{2}\bar{\frac{1}{2}}\bar{\frac{1}{2}})$ lying at the centre points of stacked cubes. In crystallography these two lattices are called primitive cubic and body centred cubic respectively. For n=2 the distinction is clearer, the primitive hypercubic lattice is generated by vertices of stacked squares, while the primitive hyperbolic lattice is generated by vertices of stacked hexagons.

(iv) It is not true that if H acts transitively on a set then G(T) acts fully transitively, although this happens to be the case for n=3.

Theorem Suppose $H \subset O(n)$ is fully transitive. Then the set of k vectors on which it acts may be chosen to be either $\{e_1, \ldots, e_n\}$ or $\{f_1, \ldots, f_{n+1}\}$ as in the above examples.

Proof By hypothesis H induces all possible permutations on a set $A = \{t_1, \ldots, t_k\}$. The vector $t_1 + \ldots + t_k$ is invariant and therefore zero. If $t_i \in A \Rightarrow \bar{t}_i \in A$ for a particular value of i, then the same must be true for all i. In this case $A = \{\pm, \ldots, \pm e_n\}$ because the symmetry which reverses e_i and leaves all other e_j fixed shows that e_i, e_j are perpendicular. If on the other hand $\bar{t}_i \notin A$ for $i=1, \ldots, n$ then $k = n+1$ since $k > n+1$ would imply the existence of k further linear equations relating $t_1, \ldots, t_k$.

Remark The above result is due to Hermann [24]. It implies that, if $H \subset O(n)$ is fully transitive, an arithmetic crystal class (T,H) must belong to one of the two crystal families described in (i) and (ii) above. It does not however necessarily belong to the same Bravais type or crystal system since there are other choices of T consistent with the action of H and some of these choices may lead to different symmetry groups G(T). See further 5.6 below.

4.5. ORTHOGONAL POINT GROUPS

In this section we consider certain geometric crystal classes which are of importance because the number of associated space groups is very large. Let $e_1,\ldots,e_n$ be mutually orthogonal vectors in R^n, not necessarily of equal length, and let μ_i be the reflection defined by

$$\mu_i(e_j) = \begin{cases} e_j, & j \neq i \\ -e_j, & j = i \end{cases}$$

We write $\bar{\mu}_i$ for the composite $\bar{1}\mu_i$ mapping e_j to $-e_j$ for $j \neq i$ and leaving e_i fixed.

Definition $H \subset O(n)$ is orthogonal if it is generated by $\theta_1,\ldots,\theta_n$ where each θ_i is either μ_i or $\bar{\mu}_i$.

If $v \in V$ then $v-\mu_i v = v+\bar{\mu}_i v$ is always a multiple of e_i. Therefore any lattice, on which H acts, contains a sublattice which (after altering the vectors e_i by scalar multiples as necessary) has the form

$$T = \{r_1e_1+\ldots+r_ne_n \text{ ; } r_i \text{ integers}\} .$$

T is minimal for the action of H if the vectors e_i all have different length; then G(T) is the group of order 2^n generated by $\mu_1,\ldots,\mu_n$.

Definition $H \subset O(n)$ is orthogonal with additional symmetry if it is generated by $\theta_1,\ldots,\theta_n$, where θ_i is μ_i or $\bar{\mu}_i$, and by further symmetries θ defined by coordinate permutations satisfying $\theta(e_i) = \pm e_{\sigma(i)}$ for some $\sigma \in S_n$.

If H is orthogonal with additional symmetry then there is an exact sequence $1 \to H_o \to H \to A \to 1$ where H_o is orthogonal and $A \subset S_n$ is a group of permutations. The group A determines a partition $n = n_1 + \ldots + n_s$ by the equivalence relation $i \sim j$ if $\alpha(i) = j$ for some $\alpha \in A$, i.e. if $\theta(e_i) = \pm e_j$ for some $\theta \in H$. The lattice T is minimal for the action of H if the vectors e_i, e_j have the same length precisely when $i \sim j$. Then G(T) is generated by $\mu_1, \ldots, \mu_n$ and all permutation symmetries $e_i \longmapsto e_{\sigma(i)}$ where $\sigma \in S_{n_1} \times \ldots \times S_{n_s}$. We have

<u>Proposition</u> There are $\pi(n)$ crystal families for which H is orthogonal with additional symmetry.

<u>Remarks</u> (i) This is not the collection of $\pi(n)$ crystal families mentioned in the example of 4.4; but the orthogonal family with no additional symmetry does occur in both lists.

(ii) When n=2 the families corresponding to (2) and (1+1) are called *square* and *rectangular* respectively. For n=3 the usual names are (3) *cubic*, (2+1) *tetragonal*, (1+1+1) *orthorhombic*. For n=4 the names given by Bülow, Neubüser and Wondratschek [10] are (4) *hypercubic*, (3+1) *cubic orthogonal*, (2+2) *ditetragonal orthogonal*, (2+1+1) *tetragonal orthogonal*, (1+1+1+1) *orthogonal*.

<u>Proposition</u> The number of orthogonal geometric crystal classes is $[\frac{n}{2}]+1$.

<u>Proof</u> By definition an orthogonal group H_o is generated by n elements $\theta_1, \ldots, \theta_n$. Now $\mu_1 \ldots \mu_n = \bar{\iota}$, and therefore we may assume that $\theta_i = \mu_i$ for all except possibly an *odd* number of values of i. Re-ordering the coordinate axes suitably we can assume H_o is generated by $\bar{\mu}_1, \ldots, \bar{\mu}_j, \mu_{j+1}, \ldots, \mu_n$ where

either $j = 0$ and H_o has order 2^n

or j is odd, H_o has order 2^{n-1} and the generators satisfy

$\theta_1 \ldots \theta_n = \bar{\mu}_1 \ldots \bar{\mu}_j \mu_{j+1} \ldots \mu_n = \bar{\iota}\mu_1 \ldots \mu_n = \iota$.

There is one crystal class of the first kind, denoted m....m, and $\frac{n}{2}$

crystal classes of the second kind, denoted 2....2m....m.

Remark The geometric crystal classes arising in this way for low dimensions are: (n=2) mm,2m; (n=3) mmm,2mm; (n=4) mmmm, 2mmm, 222m. This is a slight variation on the usual Hermann-Mauguin notation but consistent with 4.1 and 4.2.

5 Lattices

The determination by Bravais [6] of the 14 lattice types in 3-dimensional space was closely related to the contemporary work of Dirichlet on quadratic forms. In this chapter we derive all arithmetic crystal classes for $n = 2,3$ by using methods from the theory of quadratic forms [56]. For arbitrary dimensions it is natural to consider particular families of lattices; this approach was initiated by Hermann [24] and continued in [48], [27].

5.1. REDUCED BASES

By 3.1, if $T \subset R^n$ is a discrete subgroup with integral basis $\{t_1,\ldots,t_r\}$ and $V \subset R^n$ is a subspace then $T \cap V$ is also a discrete subgroup.

Definition A set $\{t_1,\ldots,t_k\}$ of vectors in T is *primitive* if it forms an integral basis for $T \cap V$ where V is the subspace spanned by $t_1,\ldots,t_k$. A set $\{t_1,\ldots,t_r\}$ in T is a *reduced basis* if, for all $k = 1,\ldots,r$, (i) the set $\{t_1,\ldots,t_k\}$ is primitive, and (ii) $\{t_1,\ldots,t_{k-1},s\}$ primitive implies $|t_k| \le |s|$.

Proposition Let T be the lattice of a space group. It admits a reduced basis.

Proof By 3.1 there is a non-zero vector $t_1 \in T$ of minimum length. We construct a basis $\{t_1,\ldots,t_n\}$ by induction. Suppose $\{t_1,\ldots,t_{i-1}\}$ is given and spans a vector space V. If there exists a vector $s \in T$ not in V then it can be written $s = u + w$, $u \in V^\perp$, $w \in V$. There are, among all such s, those for which $|u|$ is a minimum. Therefore the subset

$$T_o = \{s \in T;\ \{t_1,\ldots,t_{i-1},s\} \text{ is primitive}\}$$

is non-empty. Choose t_i to be a vector in T_o of minimum length.

If $\{t_1,\ldots t_r\}$ is a reduced basis, the number $s_{ij} = (t_i,t_j)$ must satisfy various inequalities. If $\sum_{i=1}^{k-1} a_i t_i$ then $(t_k,t_k) \le (t_k-t,t_k-t)$ and so $2(t_k,t) \le (t,t)$. In particular $2(t_k,t_j) \le (t_j,t_j)$ for $j < k$. This proves

<u>Proposition</u> Let $\{t_1,t_2,\ldots,t_r\}$ be a reduced basis. The numbers $s_{ij} = (t_i,t_j)$ satisfy $\pm 2s_{jk} \le s_{jj} \le s_{kk}$ for $j < k$.

<u>Examples</u> (i) If n=2 we may choose t_2 so that $s_{12} \ge 0$. Then $0 \le 2s_{12} \le s_{11} \le s_{22}$ as in 1.5. If n=3 we may choose t_2,t_3 so that $s_{12} \ge 0$, $s_{13} \ge 0$ and obtain

$$0 \le 2s_{12} \le s_{11} \le s_{22} \le s_{33}$$

$$0 \le 2s_{13} \le s_{11}$$

$$\pm 2s_{23} \le s_{22}.$$

In these cases the reduced basis has the property: if $t_1,t_2,\ldots,t_{i-1}$ span V then t_i is of minimum length among vectors in T but not in V. For suppose t_i' is not in V and $|t_i'| \le |t_i|$. We write $t_i = u+w$, $t_i' = u'+w'$ where $u,u' \in V^{\perp}$ and $w,w' \in V$, and suppose

$$w = \sum_{j-1}^{i-1} x_j t_j \qquad \text{where } |x_j| \le \tfrac{1}{2}.$$

By construction $|u| \le |u'|$. If $|u'| = |u|$ then by construction $|t_i| \le |t_i'|$. If $|u'| > |u|$ then $|u'| \ge 2|u|$and $|u'|^2 \ge 4|u|^2$. But the hypotheses imply

$$|u|^2 = |t_i|^2 - |w|^2 \ge |t_i|^2 - \sum_{j=1}^{i-1} x_j^2|t_j|^2 \ge |t_i|^2 - \frac{(i-1)}{4}|t_i|^2$$

and hence $|u'| > |u|$ gives

$$|u'|^2 \ge 4|u|^2 \ge (4-i)|t_i|^2 \ge (5-i)|t_i'|^2$$

which is a contradiction if i=1,2,3.

(ii) If n=4 then the above argument shows that we may assume that t_4 is of minimum length among vectors in T but not in the space spanned by t_1,t_2,t_3. However there may be other vectors t_4' which have the same length and which would not give a reduced basis. For example consider the lattice with reduced basis $(1000)(0100)(0010)(\frac{1}{2}\frac{1}{2}\frac{1}{2}\frac{1}{2})$. This lattice contains $t_4' = (0001)$ and $t_4 = (\frac{1}{2}\frac{1}{2}\frac{1}{2}\frac{1}{2})$. Moreover $|t_4'| = |t_4|$ but t_4' is unsuitable for a reduced basis.

(iii) If $n \geq 5$ then it is not possible in general to assume that t_5 is of minimum length among vectors not in the space spanned by t_1,t_2,t_3,t_4. For example the lattice with reduced basis $(10000)(01000)(00100)(00010)(\frac{1}{2}\frac{1}{2}\frac{1}{2}\frac{1}{2}\frac{1}{2})$ contains the vector $t_5' = (00001)$, and so contains a set of five linearly independent vectors of length 1. But any reduced basis must contain a vector of length $\sqrt{5}/2$.

5.2. ARITHMETIC CRYSTAL CLASSES FOR n=2

In practice it is convenient to adapt the notion of reduced basis by taking into account the symmetry present in a given problem. For example if n=3 and if it is known that reflection in a plane V, or rotation about an axis perpendicular to V, act on T, then it is natural to consider a basis of the form $\{t_1,t_2,t_3\}$ where t_1,t_2 lie in V. The enumeration of arithmetic crystal classes for n=3 is then obtained as a deduction from the enumeration already completed for n=2 in 1.4. The essential features of that enumeration can be summarised as follows.

By 1.4 (or 1.5) the possible symmetry groups G(T) of 2-dimensional lattices T are (figures in brackets refer to the relevant partition of 3.1): 2, mm (1+1), 4m, 6m (2). Note that G(T) always contains a rotation of order 2, so that the first case is that of minimum symmetry. There are reduced bases $\{t_1,t_2\}$ and any such is consistent with such symmetry. Hence the name *primitive* and the notation (P,2) for (T,G(T)) in this case. The only possible subgroup is $1 \subset 2$ giving the actions (T,H)

(P,2) with subgroup (P,1)

In the other reducible case G(T) = mm. The restrictions of T to the mutually perpendicular mirror lines give lattices T_1,T_2. The projections π_1, π_2 onto these lines give possibly larger lattices π_1T, π_2T : either $\pi_1T = T_1$, $\pi_2T = T_2$ or $T_1 \subset \pi_1T$, $T_2 \subset \pi_2T$ are black and white pairs of 1-dimensional lattices. There are therefore only two possibilities, denoted P and C:

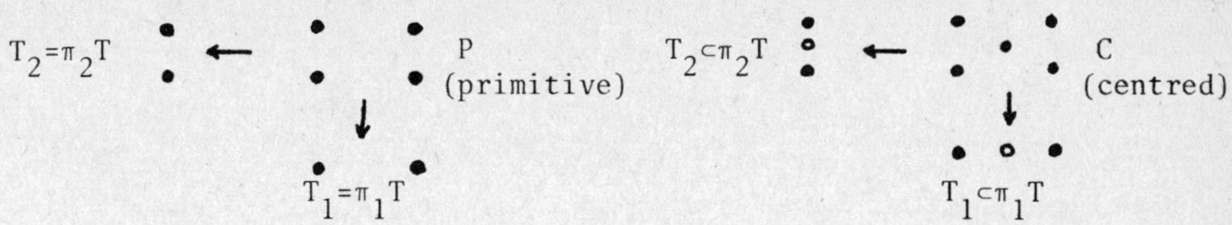

We obtain

(P, mm) with subgroup (P,m)
(C, mm) " " (C,m)

In the two irreducible cases the only possibility is for T to be generated by t and θt. This case also is called primitive, giving

(P, 4m) with subgroup (P,4)
(P, 6m) with subgroups (P,6)(P,3)(P31m)(P3m1)

In the notation for the point group the first symbol refers to the axis of rotation, the second to an axis *perpendicular to* a vector $t \in T$ of minimum length, the third to an axis in the plane different from the second (i.e. at angle $\frac{1}{2}\theta$) The reasons for these choices will become clearer in the case n=3 for which, of course, the notation was originally designed.

The listings of subgroups include only those actions for which T is minimal. There is therefore one representative for each arithmetic crystal class, giving 13 arithmetic crystal classes. Note that, in accordance with 3.4, these can be interpreted as the 13 isomorphism classes of plane groups which contain no glide transformations (other than composites of translations and reflections already in the group).

5.3. ARITHMETIC CRYSTAL CLASSES FOR n=3

Every lattice admits the symmetry $\bar{1}$. Thus we obtain

$$(P, \bar{1}) \quad \text{with subgroup} \quad (P,1) \tag{1-2}$$

Any further symmetry must include a rotation or reflection. Since $\rho\bar{\iota}$ is a reflection whenever ρ is a rotation of order 2, there are two cases: G(T) contains no reflection in which case it would have to be the group $\bar{3}$, or G(T) contains a reflection. We deal with the first case easily. Let V be the plane perpendicular to the rotational axis of order 3. Then $T \cap V$ has basis $\{t,\theta t\}$ where $\theta^3 = \iota$. There is a basis for T of the form $\{t,\theta t,w+ye_3\}$ where $w \in V$ and $e_3 \in V^\perp$ is a vector of length 1. Then

$$(\iota+\theta+\theta^2)(w+ye_3) = 3y\, e_3 \in T$$

and $3w \in T \cap V$. The positions of w must be as indicated

The symmetry group of T is therefore $\bar{3}m$: it does in fact contain a reflection but in a line through, not perpendicular to, the rotational axis. We call this lattice *rhombohedral* and have

$$(R,\bar{3}m) \quad \text{with subgroups} \quad (R,\bar{3}),(R,3m)(R,32),(R,3). \tag{3-7}$$

For any other 3-dimensional lattice T we may assume that the symmetry group G(T) contains reflections in planes perpendicular to every rotational axis which occurs. Let V be the plane perpendicular to the rotational axis of highest order, and $\{t_1,t_2\}$ a reduced basis for $T \cap V$. Denote by $\mu \in G(T)$ the reflection in V and by π the orthogonal projection on to V. There is a basis for T of the form $\{t_1,t_2\ w+ye_3\}$ where $w \in V$ and $e_3 \in V$ is a vector of length 1. Then

$$(\iota-\mu)(w+ye_3) = 2y\, e_3 \in T$$

and $2w \in T \cap V$. Therefore either $\pi T = T \cap V$ (the primitive case, denoted P) or $T \cap V \subset \pi T$ is a black and white lattice (the body centred case usually denoted I). We list the cases which arise, numbering the group actions which arise, according to the form of the 2-dimensional lattice $T \cap V$. The first column refers to the group action $(T \cap V, G(T \cap V))$

(P,2): one black and white lattice giving T = I(for *innenzentriert*)

(P,2/m)	with subgroups	(P,m),(P,2).	(8-10)
(I,2/m)	" "	(I,m),(I,2).	(11-13)

(P,mm): two black and white lattices giving T = A,I

(P,mmm)	" "	(P,2mm),(P,222).	(14-16)
(A,mmm)	" "	(A,2mm),(A,222).	(17)
(I,mmm)	" "	(I,2mm),(I,222).	(18-20)

(C,mm): one black and white lattice giving T = F (for *face centred*). The case $T \cap V = \pi T$ now gives T = C. The group actions (A,mmm),(A,222) have not been numbered because they are equivalent to (C,mmm),(C,222) under a permutation of axes (the letters A,B,C refer to the existence in the lattice of vectors $(0\frac{1}{2}\frac{1}{2})(\frac{1}{2}0\frac{1}{2})(\frac{1}{2}\frac{1}{2}0)$ respectively). We have

(C,mmm)	" "	(C,2mm),(C,222).	(21-23)
(F,mmm)	" "	(F,2mm),(F,222).	(24-26)

(P,4m): one black and white lattice giving T = I. Now 4/m mm has many subgroups which must be listed with due regard to the action on T. Thus $\bar{4}2m$ has rotation of order 2 with axis in $T \cap V$ along a shortest vector, and reflection in a plane perpendicular to a diagonal axis in $T \cap V$. On the other hand $\bar{4}m2$ has reflection in a plane perpendicular to a shortest vector, and rotation about a diagonal axis in $T \cap V$. Thus we obtain

(P,4/mmm)	" "	(P,$\bar{4}$2m),(P,$\bar{4}$m2),(P,4/m),(P,4mm), (P,422),(P,4),(P,$\bar{4}$).	(27-34)
(I,4/mmm)	" "	(I,$\bar{4}$2m),(I,$\bar{4}$m2),(I,4/m)(I,4mm), (I,422),(I,4),(I,$\bar{4}$)	(35-42)

(P,6m): no black and white lattice but many subgroups of 6/m mm giving

(P,6/m mm)	" "	(P,$\bar{6}$2m),(P,$\bar{6}$m2),(P,6/m),(P,6m), (P,622),(P,6),(P,$\bar{6}$),(P3),(P,$\bar{3}$),(P,$\bar{3}$m1), (P,$\bar{3}$1m),(P,321),(P,312),(P,3m1),(P,31m).	(43-58)

In all the above cases it has been assumed that T is minimal; that is the lattice parameter y is chosen in general position rather than with any special value giving higher symmetry. Such special values correspond to symmetry groups with partition 3 rather than, as above, 1+1+1 and 2+1. For example (P,4/mmm) with $y = \frac{1}{2}|t_1|$, and (I,4/m mm) with $y = \frac{1}{2}|t_1|$ or $y = 1/\sqrt{2}|t_1|$ give various cubic lattices

(P,m3m) " " (P,$\bar{4}$3m),(P,432),(P,m3),(P,23). (59-63)

(I,m3m) " " (I,$\bar{4}$3m),(I,432),(I,m3),(I,23). (64-68)

(F,m3m) " " (F,$\bar{4}$3m),(F,432),(F,m3),(F,23).. (69-73)

The same arise by specialisation from (R,$\bar{3}$m) by choosing $y = \sqrt{1/6}|t_1|$, $\frac{1}{2}\sqrt{1/6}|t_1|$, $\sqrt{2/3}|t_1|$. To be sure that the 73 arithmetic classes give a complete list for n=3 it is necessary to check that no further special values occur. Since specialisation must increase the symmetry group. But the only inclusions between symmetry groups are those indicated in the diagram of subgroup/supergroup relationships:

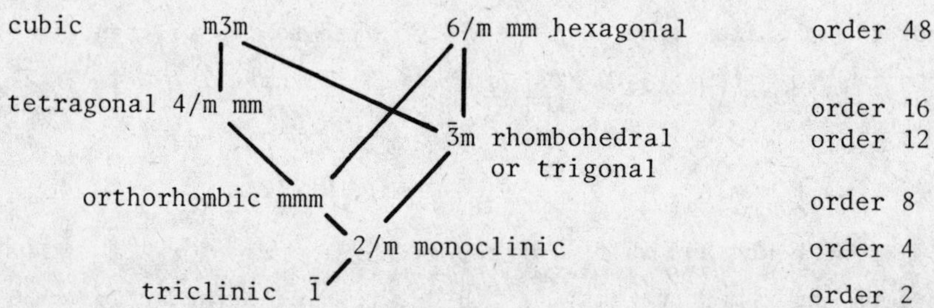

5.4. HERMANN-MAUGUIN NOTATION, SYSTEMS AND FAMILIES, FOR n=3

The listing of the 73 arithmetic crystal classes given in 5.3 should make clear how the Hermann-Mauguin notation is used. The symbols in successive positions refer to (i) Bravais type of lattice, (ii) rotations about, or reflections in plane perpendicular to, vertical or z axis, (iii) rotations about, or reflections in plane perpendicular to, the *first* horizontal axis, (iv) some for the *second* non-vertical axis, or, if inappropriate, some other axis. The meaning of (iii) and (iv) will vary with the situation under discussion, the intention being that neither should ever be a redundant axis whose behaviour can be deduced from those going before.

One advantage of the notation is that in cases where a rotation or reflection appears in a space group (through combination with a translation)

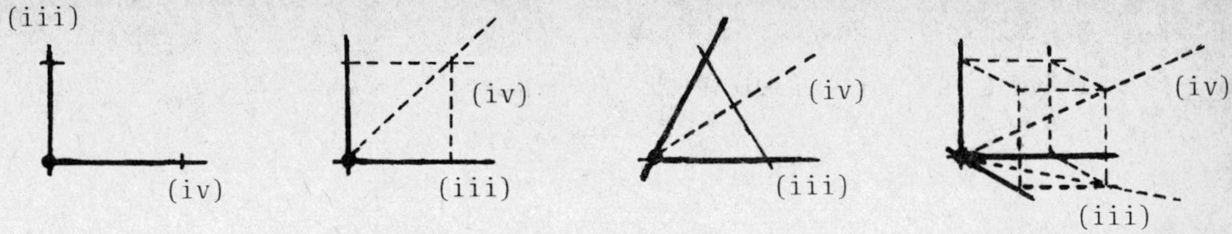

mmm 4/mmm $6/m\ mm$ or $\bar{3}m$ m3m

as a screw or glide transformation the appropriate symbol can be altered to indicate this: for example 3_1 for a screw transformation with rotation part of order 3, or g replacing m to indicate a glide transformation. The notation is therefore well adapted to listing space groups. Similarly it can be used for black and white groups. One defect, however, is that it does not give a natural listing of lattice types, each of which is given a name ad hoc. In 5.5 we show how to give a natural listing of a large class of lattices.

Note that the Hermann-Mauguin notation gives instant recognition of the various crystal families for n=3. Thus $(R,\bar{3}m)$ belongs to the same family as (P,6/m mm) because there are groups H, namely $\bar{3}m$ and its subgroups $3m, 32, \bar{3}, 3$ for which (P,H) and (R,H) are both minimal lattices. It is also clear how to assign arithmetic crystal classes to crystal systems: thus (P,3) is assigned to 6/m mm while (R,3) is assigned to $\bar{3}m$. However it is less clear what is the natural way to assign a geometric crystal class to a crystal system: if 3 is assigned to $\bar{3}m$ on the grounds that $R \leq P$ then this assignation is not consistent with that of the arithmetic crystal class (P,3).

5.5. ORTHOGONAL LATTICES

The Bravais types which belong to the orthogonal family are particularly numerous and can be described easily. In fact they have been studied independently in linear algebra and in coding theory. We adopt the notation of 4.5, where μ_i is reflection in the hyperplane perpendicular to $e_i \in R^n$ and where $e_1,\ldots,e_n$ are orthogonal but not necessarily of the same length.

Definition T is *orthogonal* if G(T) is the group of order 2^n generated by $\mu_1,\ldots,\mu_n$. T is *orthogonal with additional symmetry* if it is generated by

$\mu_1,\ldots,\mu_n$ and permutation transformations $e_i \longmapsto \pm e_{\alpha(i)}$ for $\alpha \in A \subseteq S_n$.

If T is orthogonal then, by an argument used already in 4.5, it contains non-zero vectors along the axis e_i. It is convenient to choose scalar multiples so that $\pm 2e_i$ are of minimum length among such vectors. Then any $t \in T$ has $t \pm \mu_i t \in T$ and hence t is an integral combination of $e_1,\ldots,e_n$. Let k be the field {0,1} and define a subspace $W \subset k^n$ by

$$(c_1,\ldots c_n) \in W \Longleftrightarrow \sum_{r=i}^{n} c_i e_i \in T.$$

Any lattice with the same Bravais type as T will give the same subspace W up to permutations of the coordinates. The following result is now immediate.

<u>Theorem</u> The Bravais types of orthogonal lattice T in dimension n are in one-one correspondence with equivalence classes of subspaces $W \subset k^n$ containing no basis vectors. Two subspaces W,W' are equivalent if $W' = \sigma W$ for some $\sigma \in S_n$ acting on coordinate axes.

<u>Examples</u> The correspondence for the lattices already identified when $n \leq 3$ is

n = 1	T = (0).
n = 2	P = (00); C = (00)(11).
n = 3	P = (000); C = (000)(110); I = (000)(111);
	F = (000)(011)(101)(110).

For $n \leq 5$ the number of equivalence classes is 2^{n-1}. Thus for n=4 there are 8 orthogonal Bravais types (the symbols and names given below are translations of those given by Bülow, Neubüser and Wondratschek [10]):

P = (0000) primitive; S = (0000)(1100) side face centred; I = (0000)(1110) body centred; Z = (0000)(1111) centrally centred; F = (0000)(0110)(1010)(1100) face centred; D = (0000)(1100)(0011)(1111) double side face centred; G = (0000)(0111)(1011)(1100) mixed centred; U = (0000)(1100)(1010)(1001)(0110)(0101)(0011)(1111) everywhere side face centred.

However for n=6,7 there are 36,80 equivalence classes of subspaces, the behaviour for larger n being given by:

<u>Proposition</u> Let 0_n be the number of Bravais types of orthogonal lattice in dimension n. Then $0_n/2^{n-1} \to \infty$ as $n \to \infty$.

<u>Proof</u> Let $d_{r,n}$ denote the number of r-dimensional subspaces $W \subset k^n$ which contain no basis vector. Thus $d_{o,n} = 1$, $d_{n-1,n} = 1$ and $d_{1,n} = 2^n-(n+1)$. Clearly for $1 < r < n$

$$d_{r,n} > \frac{(2^n-1)(2^{n-1}-1)\dots(2^{n-r+1}-1)}{(2^r-1)(2^{r-1}-1)\dots(2-1)} - n\,\frac{(2^{n-1}-1)\dots(2^{n-r+1}-1)}{(2^{r-1}-1)\dots(2-1)}$$

$$= \left(\frac{2^n-1}{2^r-1} - n\right)\prod_{s=1}^{r-1}\frac{(2^{n-r+s}-1)}{(2^s-1)}$$

$$> \left(\frac{2^n-1}{2^r-1} - n\right)\ 2^{(n-r)(r-1)}$$

Therefore $d_{r,n} > 2^{(n-r)r-1}$ whenever the inequality

$$\frac{2^n-1}{2^r-1} - n > 2^{n-r-1}$$

holds. Now for sufficiently large n = 2m or 2m+1 the inequality will hold for r between m-2 and m+1. Then

$$\sum_r d_{r,n} > \sum_{r=m-2}^{m+1} 2^{(n-r)r-1} > \begin{cases} (2^{m^2-1}+2^{m^2}+2^{m^2}+2^{m^2-1}) = 2^{m^2} & \text{if } n = 2m \\ (2^{m^2+m-2}+2^{m^2+m}+2^{m^2+m}+2^{m^2+m-2}) = \frac{5}{4}\,2^{m^2+m} & \text{if } n = 2m+1 \end{cases}$$

Since $\frac{5}{4} > 2^{\frac{1}{4}}$ we always have

$$\sum_r d_{r,n} > 2^{\frac{1}{4}n^2},$$

and hence

$$O_n > 2^{\frac{1}{4}n^2}/n!$$

The result follows from the fact that n! tends to infinity like $n \log n$ and is therefore of strictly smaller order than $2^{\frac{1}{4}n^2-n+1}$.

If now T is orthogonal with additional symmetry the same construction gives a vector subspace $W \subset k^n$ with the property: W contains no basis vectors and $\sigma W = W$ for all $\sigma \in A$ where $A \subseteq S_n$ is the subgroup defining the additional symmetry. There are two differences from the previous case: the equivalence class of W may not be defined uniquely, because there may be more than one choice of initial coordinate directions $e_1,\ldots,e_n$, and the symmetry group of T might have order larger than $2^n|A|$.

Suppose that the action of A on coordinate axes induces a partition $n_1+\ldots+n_s$ of n corresponding to blocks on which A is transitive. Then the symmetry group of T contains all coordinate transformations induced by $\sigma \in S_{n_1}\times\ldots\times S_{n_s}$. If a particular block of size $n_\ell = m$ admits alternative initial coordinate directions then we may number the axes so that the two choices are $e_1,\ldots,e_m$ and $f_1,\ldots,f_m$. Then

$$2f_i = \sum_j r_{ij}e_j \qquad 2e_j = \sum_k s_{jk}f_k$$

for some integers r_{ij}, s_{jk}. Comparison of lengths shows

$$r = \sum_j r_{ij}^2 \text{ is independent of } i,$$

$$s = \sum_k s_{jk}^2 \text{ is independent of } j,$$

$$rs = 16.$$

There are two cases to consider.

(i) r=2, s=8. Then $2f_i$ has the form $\pm e_{j_1} \pm e_{j_2}$. It follows that m is even and axes can be renumbered so that

$$2f_1 = e_1+e_2,\ 2f_2 = e_1-e_2,\dots,2f_{m-1} = e_{m-1}+e_m,\ 2f_m = e_{m-1}-e_m.$$

By transitivity T must contain all vectors of the form $\pm e_{j_1} \pm e_{j_2}$ and hence all vectors of the form $\pm f_{k_1} \pm f_{k_2} \pm f_{k_3} \pm f_{k_4}$. If $m > 4$ then T contains all $\pm f_{k_1} \pm f_{k_2}$ and so the subspace W is not altered by the new choice of initial coordinate directions. If m=4 the same lattice T arises from the subspaces (0000...0)(1111...0) and (0000...0)(1100...0)(1010...0)(1001...0)(0110...0)(0101...0)(0011...0)(1111...0) of k^n. If m=2 then the same lattice T arises from the subspaces (00...0)(11...a) and (00...0)(10...a)(01...a)(11...0) of k^n, if $n > 2$, or from the subspaces (00) and (00)(11) if n=2.

(ii) r=4, s=4. Then $2f_i$ has the form $\pm e_{j_1} \pm e_{j_2} \pm e_{j_3} \pm e_{j_4}$. Renumber so that $2f_1, 2f_2, 2f_3, 2f_4$ is one of the sets expressed in terms of the original basis e_1, e_2, e_3, e_4 as $\{(\bar{1}111)(1\bar{1}11)(11\bar{1}1)(111\bar{1})\}$ or $\{(\bar{1}\bar{1}11)(\bar{1}1\bar{1}1)(1\bar{1}\bar{1}1)(1111)\}$. If $m > 4$ then transitivity implies that T contains all vectors of the form $\pm e_{j_1} \pm e_{j_2}$, and hence also of the form $\pm f_{k_1} \pm f_{k_2}$, which leads quickly to a contradiction. If m=4 then the lattice arising from the subspace (0000...0)(1111...0) has symmetry group of order $4!2^4.3 = 1152$, the scalar factor 3 corresponding to even permutations of the three choices $\{(2000)(0200)(0020)(0002)\}$, $\{(\bar{1}111)(1\bar{1}11)(11\bar{1}1)(111\bar{1})\}$, $\{(\bar{1}\bar{1}11)(\bar{1}1\bar{1}1)(1\bar{1}\bar{1}1)(1111)\}$ for $2e_1, 2e_2, 2e_3, 2e_4$. The subspace $W \subset k^n$ must in this case have a direct summand (0000...0)(1111...0) or equivalently (0000...0)(1100...0)(1010...0)(1001...0)(0110...0)(0101...0)(0011...0)(1111...0).

Examples (i) If n=3 and $n_1=2$, $n_2=1$ there are equivalences between the subspaces (000) and (000)(110), and also between (000)(111) and (000)(110)(101)(011). This explains why the four orthorhombic Bravais types P,C,I,F of 5.3 specialise to only two tetragonal types: there are *lattice coincidences* $P \sim C$ and $I \sim F$.

(ii) If n=4 and $n_1=n_2=2$ the only subspaces $W \subset k^4$ invariant under $S_2 \times S_2$ and containing no basis vector are those corresponding to the orthogonal Bravais types P,S,Z,D,G,U described above. However there are lattice coincidences $P \sim S \sim D$ and $Z \sim G \sim U$ and so only two ditetragonal orthogonal Bravais types.

(iii) If $n=n_1=4$ the only subspaces $W \subset k^4$ invariant under S_4 and containing no basis vector are those corresponding to the orthogonal Bravais types P,Z,U. There are lattice coincidences $Z \sim U$ as above giving two hypercubic Bravais types with symmetry groups order 384, 1152 respectively.

In the following propositions and theorem $W \subset k^n$ is a subspace invariant under $S_{n_1} \times \ldots \times S_{n_s}$; the map $\pi_i : k^n \to k^{n_i}$ is projection on to a particular block i of coordinate axes; e_j denotes the j-th basis vector of k^n as well as the corresponding vector in $\frac{1}{2}T$, while e_i denotes the i-th basis vector of k^s.

Proposition If $n = n_1 > 4$ then the only W invariant under S_n and containing no basis vector are $P_{(n)} = (0\ldots0)$, $Z_{(n)} = (0\ldots0)(1\ldots1)$ and $U_{(n)} = \{(r_1\ldots r_n);\ \Sigma_j r_j = 0\}$. There are three hypercubic Bravais types each with symmetry group order $n!\ 2^n$.

Proof Since $n > 4$ there is no possibility of any lattice coincidence between $Z_{(n)}$ and $U_{(n)}$.

Proposition If $\pi_i W$ contains a vector of the form $e_{j_1}+e_{j_2}$ and $n_i > 2$ then W has a direct summand $U_{(n_i)}$.

Proof Since $n_i > 2$ the space W contains a vector (a...110...b). By transitivity it also contains (a...101...b) and hence (0...011...0). Now transitivity implies that W contains $U_{(n)}$ as a direct summand.

Theorem The Bravais types of lattices T in dimension n which are orthogonal with additional symmetry $A \subseteq S_n$ are determined by:

(i) a partition π of n into blocks of axes on which A acts transitively, $n = n_1+\ldots+n_s$;

(ii) a list L of blocks i all with $n_i \neq 1,2,4$;

(iii) a vector subspace $E \subseteq k^s$ which can contain e_i only if $n_i \neq 1,2$ or $i \in L$.

The Bravais types determined by π,L,E and π',L',E' coincide if there is a permutation $\sigma \in S_s$ such that $n_i' = n_{\sigma(i)}$, $L' = \sigma L$, $E' = \sigma E$.

Proof An orthogonal lattice T with additional symmetry $A \subseteq S_n$ is determined by a subspace $W \subset k^n$ invariant under $S_{n_1} \times \ldots \times S_{n_s}$. By the previous proposition

$$W = \bigoplus_{i \in L} U_{(n_i)} \oplus V$$

for some list L with $n_i \neq 1,2,4$ for $i \in L$ because of lattice coincidences. If $i \notin L$ then $\pi_i V$, being invariant under S_{n_i}, must be either $P_{(n_i)}$ or $Z_{(n_i)}$. If $i \in L$ then $\pi_i V$ can be chosen to be $(0\ldots0)$ or $(0\ldots0(r_1\ldots r_n)$ for $r_j \in Z$ with $\sum_j r_j = 1$. Define the vector space $E \subseteq k^s$ by contraction of coordinates of V; that is, $E = \alpha V$ where αv has i-th coordinate 0 for $\pi_i v = 0$ and i-th coordinate 1 for $\pi_i v = (1\ldots1)$ if $i \notin L$ or $\pi_i v = (r_1 \ldots r_n)$ with $\sum_j r_j = 1$ if $i \in L$. There is an inverse map $\delta : k^s \to k^n$ defined by $\delta e_i = (0\ldots011\ldots110\ldots0)$ for $i \notin L$, such that $\delta E = V$. The condition $n_i \neq 1,2$ for $e_i \in E$ follows because of lattice coincidences and because W does not contain any basis vectors; similarly $i \notin L$ for $e_i \in E$. Finally any permutation in S_n, which is consistent with the partitions π, π' and maps W to W', must induce a permutation $\sigma \in S_s$ as specified.

5.6. LATTICES WITH HIGH SYMMETRY

Among the lattices defined for arbitrary values of n are several related to the hypercubic lattice P and the hyperbolic lattice H (see 4.4). We adopt the conventions that $i \neq j$, that P has integral generators $e_1, \ldots, e_n$ satisfying $(e_i, e_i) = 1$, $(e_i, e_j) = 0$, and that H has integral generators $f_1, \ldots, f_{n+1}$ satisfying $\sum_i f_i = 0$, $(f_i, f_i) = 1$, $(f_i, f_j) = -\frac{1}{n}$. The lattices related to P are:

- Z generated by $2e_i, e_1 + \ldots + e_n$ for $i = 1, \ldots, n$,
- U generated by $e_i \pm e_j$ for $i,j = 1, \ldots, n$,
- Q generated by $e_i \pm e_j$, $\frac{1}{2}(e_1 + \ldots + e_n)$ for $i,j = 1, \ldots, n$, and defined whenever n is even.

They satisfy the relationships $2P \subset Z$, $2P \subset U \subset Q$, and if n is even $Z \subset U$, and there are special coincidences:

n = 2: Q primitive orthogonal with symmetry group 2mm

P has same Bravais type as Z = U with symmetry group 4mm.

n = 4: Q has same Bravais type as P

Z has same Bravais type as U

To prove the statements for n=4, we use the ideas of 4.4, namely to map P to Q by mapping (1000),(0100),(0010),(0001) to $(\frac{1}{2}\frac{1}{2}\frac{1}{2}\frac{1}{2})$, $(\frac{1}{2}\frac{1}{2}\frac{\bar{1}}{2}\frac{\bar{1}}{2})$, $(\frac{1}{2}\frac{\bar{1}}{2}\frac{1}{2}\frac{\bar{1}}{2})$, $(\frac{1}{2}\frac{1}{2}\frac{\bar{1}}{2}\frac{\bar{1}}{2})$, and to map Z to U by mapping (2000),(0200),(0020),(0002) to (1100),$(1\bar{1}00)$,(0011) $(001\bar{1})$. The symmetry groups of P,Z,U contain the full monomial group of order $n!2^{n-1}$ in which only even numbers of sign changes of coordinates are permitted. To see that, for almost all values of n, this is the whole symmetry group we consider the shortest non-zero vectors of the various lattices (each pair $\pm v$ is counted as a *single* shortest vector).

P	n vectors length 1, scalar products 0	
Z	n vectors length 2 2^{n-1} vectors length $\sqrt{n}$	for n=4 have 12=4+8 vectors length 2
U	$2\binom{n}{2}$ vectors length $\sqrt{2}$, scalar products 0 or ± 1.	
Q	$2\binom{n}{2}$ vectors length $\sqrt{2}$ 2^{n-2} vectors length $\frac{1}{2}\sqrt{n}$	for n=4 have 4 vectors length 1 for n=8 have 120=56+64 vectors length $\sqrt{2}$

It is now easy to check that P,Z,U do indeed have symmetry group of order $n!2^n$ except in case n=4 (when Z = U has symmetry group order 1152), and that Q has symmetry group order $n!2^{n-1}$ except in case n=4,8.

The lattice Q for n=8 has many remarkable properties. It has the same Bravais type as another lattice $S_{(3)}$ defined below and its symmetry group has order $(8!2^7)(3^3.5)$. Notice that the decomposable lattice T = Q × Q in 16-dimensional space has 240 vectors length $\sqrt{2}$, and that $240 = 2\binom{16}{2}$. Thus for n=16 the lattices T and Q have the same number of shortest non-zero vectors; it can in fact be shown that for *all* values c the number of vectors in T and Q of length c is the same. Nevertheless T and Q do not have the same Bravais type, nor even the same symmetry group.

The lattices related to H are:

$S = \{\sum_i r_i f_i \; ; \; \sum_i r_i = 0 \bmod n+1\}$ generated by $f_i - f_j$ for

$i,j = 1,\ldots,n+1$

$S_d = \{\sum_i r_i f_i \; ; \; \sum_i r_i = 0 \bmod d\}$, defined whenever d is a proper factor

of n+1, and generated by $f_{i_1} + \ldots + f_{i_d}$ for $i_1,\ldots,i_d = 1,\ldots,n+1$.

They satisfy the relationships $dh \subset S_d$, $(n+1)H \subset S \subset S_d$ and there are special coincidences:

n = 2: S has same Bravais type as H with symmetry group 6mm

n = 3: H,S,S_2 have same Bravais type as Z,U,P respectively.

The symmetry groups of H,S,S_2 all contain the group of order $(n+1)!2$ generated by permutations and simultaneous sign change of $f_1,\ldots,f_{n+1}$. To see that, for almost all values of n, the symmetry group is no larger consider the shortest non-zero vectors of the various lattices. We assume $n > 3$, $d > 2$.

H $n+1$ vectors length 1, scalar products $\frac{-1}{n}$

S $\binom{n+1}{2}$ vectors length $\sqrt{(2+\frac{2}{n})}$, scalar products 0 or $\pm(1+\frac{1}{n})$

S_2 $\binom{n+1}{2}$ vectors length $\sqrt{(2-\frac{2}{n})}$, scalar products $\pm\frac{4}{n}$ or $\pm(1-\frac{3}{n})$

In general S_d has vectors of length $\sqrt{(d - \frac{d(d-1)}{n})}$ numbering $\binom{n+1}{d}$ or $\frac{1}{2}\binom{n+1}{d}$ according as $d \neq \frac{1}{2}(n+1)$ or $d = \frac{1}{2}(n+1)$. In addition S_d contains vectors of the sublattices S and $S_{d'}$ for $d' | d$. If $d \geq 5$ then $d - \frac{d(d-1)}{n} > d - \frac{1}{2}d \geq \frac{5}{2}$ and so the shortest non-zero vectors of S_d are those of S. We therefore have

S_d	$(d \geq 5)$	$\binom{n+1}{2}$ vectors length $\sqrt{(2+\frac{2}{n})}$, scalar products 0
S_3, S_4	$(n \geq 11)$	or $\pm(1+\frac{1}{n})$
S_3	$(n=5)$	$\frac{1}{2}\binom{6}{3}=10$ vectors length $\frac{3}{\sqrt{5}}$, scalar products $\pm\frac{3}{5}$.
	$(n=8)$	$\binom{9}{3}+\binom{9}{2}=120$ vectors length $\frac{3}{2}$, scalar products 0, $\pm\frac{9}{8}$.
S_4	$(n=7)$	$\frac{1}{2}\binom{8}{4}+\binom{8}{2}=63$ vectors length $\frac{4}{\sqrt{7}}$, scalar products 0, $\pm\frac{8}{7}$.

It is now easy to check that for H,S there can be no additional symmetries; that for S_2 there can be no additional symmetries so long as $\frac{4}{n} \neq 1-\frac{3}{n}$, that is except for n=7; and that for S_d there can be no additional symmetries except for d=3, n=8 and d=4, n=7. The only lattices with order of symmetry group larger than (n+1)!2 are $S_4 \subset S_2$ in case n=7 and S_3 in case n=8. In fact the orders are $(8!2)(2^2.3^2)$ and $(9!2)(2^6.3.5)$.

Inspection of the shortest vectors and their scalar products shows that those for S_3 and Q coincide, up to a constant multiple, for n=8. These lattices do indeed have the same Bravais type.

5.7. ORTHOGONAL BLACK AND WHITE LATTICES

Recall the definition in 3.6 of a class equivalent colour symmetry group. It is a pair $G_1 \subseteq G$ of index r with $T_1 \subseteq T$ and $H_1 = H$. Thus the pair $T_1 \subseteq T$ is invariant under the action of $H \subset O(n)$. If H is generated by $\theta_1,\ldots,\theta_n$, where each θ_i is μ_i or $\bar{\mu}_i$, the pair is called *orthogonal*; if r=2 it called a *black and white lattice*. The analogues for orthogonal black and white lattices of the theorems of 5.5 are

Theorem The Bravais types of orthogonal black and white lattices in dimension n are in one-one correspondence with equivalence classes of pairs of subspaces $W_1 \subset W$ of k^n such that W_1 contains no basis vector and $|W/W_1| = 2$. Two pairs $W_1' \subset W'$, $W_1 \subset W$ are equivalent if $W_1' = \sigma W_1$, $W' = \sigma W$ for some $\sigma \in S_n$ acting on coordinate axes.

Proof Choose $e_1,\ldots,e_n$ so that $\pm 2e_i$ are the shortest non-zero vectors of T_1 along the i-th axis. By 5.5 the lattice T_1 is determined by a subspace

$W_1 \subset k^n$ which contains no basis vector. Since $T_1 \subset T$ has index 2, and T is orthogonal, any $t \in T$ must satisfy $t \pm \mu_i t \in T_1$. Therefore $t = \sum_i r_i e_i$ where r_i Z and t corresponds to a vector $w \in k^n$. Let W be the subspace of k^n spanned by w and W_1 (note that W may contain a basis vector).

<u>Theorem</u> The Bravais type of black and white lattices in dimension n which are orthogonal with additional symmetry $A \subseteq S_n$ are determined by:

(i) a partition π of n into blocks of axes on which A acts transitively, $n = n_1 + \ldots + n_s$;

(ii) a list L of blocks i, all with $n_i \neq 1,2,4$;

(iii) a vector subspace $E_1 \subseteq k^s$ which can contain e_i only if $n_i \neq 1,2$ and $i \notin L$;

(iv) a vector subspace $E \subseteq k^s$ with $E_1 \subset E$ of index 2.

The Bravais types determined by π, L, E_1, E and π', L', E_1', E' coincide if there is a permutation $\sigma \in S_s$ such that $n_i' = n_{\sigma(i)}$, $L' = \sigma L$, $E_1' = \sigma E_1$, $E' = \sigma E$.

<u>Proof</u> By 5.5 the lattice T determines π, L, E_1. Any T invariant under the action of A which contains T must correspond to a subspace $W \subseteq k^n$ containing

$$W_1 = \bigoplus_{i \in L} U_{(n_i)} \oplus \delta E_1$$

as a subspace of index 2. Since $k^{(n_i)}$ is generated by $U_{(n_i)}$ and δe_i, the space W must have the form

$$W = \bigoplus_{i \in L} U_{(n_i)} \oplus \delta E$$

for $E \subseteq k^s$ containing E_1 as a subspace of index 2.

In the orthogonal case considered above the order of H is divisible by 2^{n-1} or some higher power of 2. By way of contrast consider the case where the order of H is odd.

<u>Theorem</u> If the order of H is odd, and the action of H on R^n is irreducible, then there is no black and white lattice $T_1 \subset T$ consistent with the action of H.

<u>Proof</u> Suppose $T_1 \subset T$ of index 2 exists. Choose $t \in T\backslash T_1$. Then the vector $\sum_{\theta \in H} \theta t$ is fixed by the action of H and generates an invariant subspace in R^n. Therefore $\sum_{\theta \in H} \theta t = 0$. On the other hand $|H|$ odd implies $\sum_{\theta \in H} \theta t \in T\backslash T_1$ which gives a contradiction.

In general a black and white lattice $T_1 \subset T$ consistent with the action of H determines a H-submodule T_1' of $T/2T = (Z/2Z)^n$ of index 2. Conversely if $T_1' \subset T/2T$ has index 2 then each element of T_1' has a representative $t_1 \in T$ determined modulo 2T. The set of all such t_1 forms a sublattice $T_1 \subset T$ of index 2. This construction is a special case of that considered in the next section.

5.8. COLOUR LATTICES

A pair of lattices $T_1 \subset T$ consistent with the action of H is called a *colour lattice*, the index r being the number of colours. Let T be generated by T_1 and vectors t_α, and let m be the smallest integer such that $mt_\alpha \in T_1$ for all α (m divides r but may be smaller). Then

$$mT \subseteq T_1 \subset T.$$

<u>Theorem</u> Let me as above have prime factorisation $m = p_1^{s_1} \ldots . p_k^{s_k}$. Then there is a chain of embeddings

$$T_1 \subset T_2 \subset \ldots . \subset T_k \subset T$$

such that $p_i^{s_i} T_{i+1} \subseteq T_i \subset T_{i+1}$

<u>Proof</u> Let T_i be the lattice generated by T_1 and the vectors $p_i^{s_i} \ldots . p_{k-1}^{s_{k-1}} p_k^{s_k} t$. Then $p_i^{s_i} T_{i+1} \subseteq T_i$ by construction. Clearly $T_i \subset T_{i+1}$.

The chain of embeddings constructed in the theorem is not unique because it depends on the order in which the prime power factors $p_i^{s_i}$ occur. However we do have

Theorem The submodule

$$\frac{T_i}{p_i^{s_i} T_{i+1}} \subset \frac{T_{i+1}}{p_i^{s_i} T_{i+1}} = Z^n \otimes \frac{Z}{p_i^{s_i} Z}$$

is determined uniquely by the embedding $T_1 \subset T$.

Proof Suppose that, in some other ordering of the prime power factors, p_i occurs as the first prime. Now it determines a lattice T_2' such that $p_i^{s_i} T_2' \subseteq T_1 \subset T_2'$ where T_2' is the lattice generated by T_1 and the vectors $p_1^{s_1} \ldots p_{i-1}^{s_{2-1}} p_{i+1}^{s_{i+1}} \ldots p_k^{s_k} t_\alpha$. To simplify the notation write $u_\alpha = p_{i+1}^{s_{i+1}} \ldots . p_k^{s_k} t$. We obtain the diagram

$$\begin{array}{ccccccl} p_i^{s_i} T_{i+1} & \subseteq & T_i & \subset & T_{i+1} & & \text{generated by } T_1 \text{ and } u_\alpha \\ \cup & & \cup & & \cup & & \\ p_i^{s_i} T_2' & \subseteq & T_1 & \subset & T_2' & & \text{generated by } T_1 \text{ and } p_1^{s_1} \ldots p_{i-1}^{s_{i-1}} u_\alpha \end{array}$$

Factor our by all multiples of $p_i^{s_i}$. The diagram becomes

$$\begin{array}{ccccc} 0 & \subseteq & T_i/p_i^{s_i} T_{i+1} & \subset & T_{i+1}/p_i^{s_i} T_{i+1} \\ & & \downarrow\cup & & \downarrow\cup \\ 0 & \subseteq & T_1/p_i^{s_i} T_2' & \subset & T_2'/p_i^{s_i} T_2' \end{array}$$

where now the vertical embedding has an inverse, namely multiplication by $p_1^{s_1} \ldots p_{i-1}^{s_{i-1}}$. Therefore the subspaces are unique.

The above theorems reduce the study of colour lattices to the study of H-submodules

$$T_i' \subset \frac{T}{p_i^{s_i}}^{\,n} = Z^n \otimes (Z/p_i^{s_i} Z)$$

consistent with the action of H, and hence to the study of H-submodules

$$T_i' \subset Z^m \otimes (Z/p_i^{s_i}Z)$$

where the representation of H acting on R^m is indecomposable. There are three cases which arise, according as (i) p divides $|H|$, (ii) p prime to $|H|$ and $R^m \otimes C = C^m$ indecomposable, or (iii) p prime to $|H|$ and $R^m \otimes C = C^m$ decomposable.

(i) p divides $|H|$. A theorem of Brauer and Nesbitt [5] implies that if p^α is the highest power of p which divides $|H|$, and $p^\alpha|m$ then $R^m \otimes C = C^m$ indecomposable implies that $Z^m \otimes (Z/pZ)$ is indecomposable. In this case repetition of the argument will lead to the conclusion that $T_i' = 0$ and so $T_i = p_i^s T_{i+1}$. In general however p^α will not divide m and $R^m \otimes C$ will not be indecomposable; then there may be non-trivial examples of submodules $T_i' \subset Z^m \otimes (Z/p_i^{s_i}Z)$.

(ii) p prime to $|H|$ and $R^m \otimes C = C^m$ indecomposable. The same theorem of Brauer and Nesbitt, with $\alpha = 0$, implies $T_i' = 0$ and $T_i = p_i^s T_{i+1}$.

(iii) p prime to $|H|$ and $R^m \otimes C = C^m$ decomposable. There may be non-trivial examples of submodules $T_i' \subset Z^m \otimes (Z/p_i^{s_i}Z)$; the existence of such examples depending on number-theoretic properties of $|H|$, m and p.

<u>Examples</u> Let H be cyclic of order $q > 2$ and $m = 2$. Then the representation of H on R^2 is indecomposable but that on $R^2 \otimes C = C^2$ is decomposable. If p is prime to q we have case (iii). Considering now representations of H over Z/pZ, we note that there exist faithful irreducible H-modules of Z_p-dimension n where n is the smallest integer r such that $q|p^r-1$; in particular there exist faithful irreducible representations of dimension 1 whenever $q|p-1$. Therefore non-trivial submodules $T' \subset Z^2 \otimes (Z/pZ)$ exist for the following prime p:

(q = 3,6) p = 7,13,19,31,37,43,61,67,73,.....

(q = 4) p = 5,13,17,29,37,41,53,61,73,.....

For primes p not on this list the module $Z^2 \otimes (Z/pZ)$ is irreducible and so there is no non-trivial submodule T'. Illustrations of the simplest colour lattices arising in this way give plane patterns in which it is surprisingly difficult to observe the full symmetry.

$5T \subset T_1 \subset T$
Action of H = 4
q = 4, p = 5

```
4 5 1 2 3 4
1 2 3 4 5 1
3 4 5 1 2 3
5 1 2 3 4 5
2 3 4 5 1 2
4 5 1 2 3 4
```

$7T \subset T_1 \subset T$
Action of H = 3 or 6
q = 3 or 6, p = 7

```
              1 2 3 4 5 6 7
             3 4 5 6 7 1 2
            5 6 7 1 2 3 4 5
           7 1 2 3 4 5 6 7
          2 3 4 5 6 7 1 2 3
         4 5 6 7 1 2 3 4 5
        6 7 1 2 3 4 5 6 7 1
       1 2 3 4 5 6 7 1 2 3
```

The study of arithmetic classes in dimension n is equivalent to the study of finite subgroups of GL(n,Z); that is, of faithful integral representations of finite groups. The above results show that the study of colour lattices is intimately connected with modular representations of finite groups. Both subjects are difficult with few general results available. For particular, but incomplete, examples when n=2 see the pictures in [31] and [51]. A more comprehensive treatment of the case n=2 for small numbers of colours can be found in [28].

6 Determination of space groups

In principle the space groups in a given arithmetic class can be determined. The algorithm of Zassenhaus [61] used for n=4 by Bülow, Neubüser and Wondratschek [10], requires a large number of computations and so has yielded few qualitative results. The method of Maxwell [33], [7] using Coxeter diagrams of Lie algebras gives qualitative results for a wide range of point group H. In this chapter we obtain various results valid in arbitrary dimensions and of a more elementary nature (see also [50]).

6.1. POINT GROUPS CONTAINING CENTRAL INVERSION

Let (T,H) be a group action in a given arithmetic crystal class. The space groups belonging to the class are determined by equivalence classes (under action of the normaliser of H) of elements of $H^2(H,T) = H^1(H,R^n/T)$ by 3.4. This cohomology group is closely related to the weight group T* of T with respect to H.

Definition The weight group T* of T with respect to H is the set $\{v \in R^n;\ v-\theta v \in T \text{ for all } \theta \in H\}$.

Suppose that the cohomology $H^1(H,R^n/T)$ is known, and that $K \neq H$ is generated by $\bar{\iota}$ and H. Then the cohomology $H^1(K,R^n/T)$ can be determined using the following result due to Maxwell [33].

Proposition There is an isomorphism

$$H^1(K,R^n/T) \longrightarrow [H^1(H,R^n/T)]_2 \times (T^*/T)/2(T^*/T)$$

where T* is the weight group of T with respect to H and $[\ \ ]_2$ denotes the set of elements of order 2.

Proof Consider $v \in H^1(K,R^n/T)$ and let $\{v_\phi\}_{\phi\in K}$ be a representative cocycle.

If $x = v_{\bar\iota}$ and $\theta \in H$ then $v_{\bar\iota\theta} = v_{\theta\bar\iota}$ implies $x - v_\theta = v_\theta + \theta x$ mod T and therefore

$$x - \theta x = 2v_\theta \text{ mod } T.$$

This formula is used in two ways. Firstly v determined restriction $r(v) \in H^1(H,R^n/T)$ by $r(v)_\theta = v_\theta$ and the cocycle $\{2r(v)_\theta\}$ is cohomologous to zero. So $2r(v) = 0$. Conversely $u \in [H^1(H,R^n/T)]_2$ defines $i(u) \in H^1(K,R^n/T)$ by $i(u)_\theta = u_\theta$, $i(u)_{\bar\iota} = 0$ and $ri(u) = u$. Therefore $[H^1(H,R^n/T)]_2$ is a summand of $H^1(K,R^n/T)$. Secondly $v = u - ir(u)$ is an element with representative cocycle $v_\theta = 0$, $v_{\bar\iota} = x$ and the formula implies $x - \theta x = 0$ mod T, that is $x \in T^*/T$. Such a cocycle is a coboundary if and only if $x = 2y$ for $y \in T^*/T$ and therefore the complementary summand is $(T^*/T)/2(T^*/T)$.

Remark Note that the weight group T^* depends on the choice of H. For instance the weight group of T with respect to K is the subgroup $\{x \in T^*;\ 2x \in T\}$ of T^*.

6.2. DECOMPOSABLE POINT GROUPS

Suppose $H = H_\alpha \times H_\beta$ acts on $T \subset R^a \times R^b$, and denote the natural projections on to R^a, R^b by π_α, π_β. There are lattices T_α, T_β defines as the intersection of T with $R^a \times \{0\}$, $\{0\} \times R^b$ respectively. Clearly $T_\alpha \subset \pi_\alpha T$ and $T_\beta \subset \pi_\beta T$.

Lemma Let T_α^* be the weight group of T_α with respect to H_α. Then $\pi_\alpha T \subseteq T_\alpha^*$ and similarly $\pi_\beta T \subseteq T_\beta^*$. The weight group of T with respect to $H = H_\alpha \times H_\beta$ is $T_\alpha^* \times T_\beta^*$.

Proof If $t = (v,w) \in R^a \times R^b$ is in T, and $\theta \in H_\alpha$, then $t - \theta t = (v-\theta v, 0)$ is in T_α. Therefore $v \in T_\alpha^*$. Similarly $w \in T_\beta^*$. The same argument shows that if $u = (v,w)$ satisfies $u - \theta u \in T$ and $u - \phi u \in T$ for all $\theta \in H_\alpha$, $\phi \in H_\beta$ then $v \in T_\alpha^*$ and $w \in T_\beta^*$.

Proposition With the above notation, T determines an isomorphism $\pi_\alpha T/T_\alpha \longleftrightarrow \pi_\beta T/T_\beta$ between sets on which H_α, H_β act trivially.

Proof If $v \in \pi_\alpha T$ then $(v,w) \in T$ for some $w \in \pi_\beta T$. The corresponding maps $\pi_\alpha T/T_\alpha \longleftrightarrow \pi_\beta T/T_\beta$ are well-defined because $(v,w) \in T$, $(v,w') \in T$ implies $w-w' \in T_\beta$ and similarly $(v,w) \in T$, $(v',w) \in T$ implies $v-v' \in T_\alpha$. By the lemma $\pi_\alpha T/T_\alpha \subset T_\alpha{}^*/T_\alpha$ and therefore the action of H_α on $\pi_\alpha T/T_\alpha$ is trivial. Similarly the action of H_β on $\pi_\beta T/T_\beta$ is trivial.

The above proposition shows that to find lattices T with decomposable point group $H_\alpha \times H_\beta$ it is sufficient, given T_α and T_β, to determine all colour lattices $T_\alpha \subseteq \pi_\alpha T$ and $T_\beta \subseteq \pi_\beta T$ for which (i) $\pi_\alpha T \subseteq T_\alpha{}^*$ and $\pi_\beta T \subseteq T_\beta{}^*$ and (ii) the quotient groups are isomorphic.

We now suppose that $T \subset R^a \times R^b$ is given and ask how to determine the space groups G in the arithmetic crystal class determined by (T,H).

Let G be a space group with lattice T and point group $H = H_\alpha \times H_\beta$ acting on $R^a \times R^b$. We assume T_α, T_β span R^a, R^b respectively (if not then $\pi_\alpha T$, $\pi_\beta T$ must each contain directions on which H_α, H_β act trivially, and so the decomposition $R^n = R^a \times R^b$ can be altered to make the assumption true). Identifying points of $H_\alpha \times \{\iota\}$, $\{\iota\} \times H_\beta$ with H_α, H_β we may write elements of G in the form $(v,w,\theta\phi) = (v,w,\phi\theta)$ where $v \in R^a$, $w \in R^b$, $\theta \in H_\alpha$, $\phi \in H_\beta$ and so G determines the colour symmetry group $G_{1,\alpha} \subseteq G_{2,\beta}$ in R^a defined by

$$G_{1,\alpha} = \{(v,\theta);\ \theta \in H_\alpha \text{ and } (v,0,\theta) \in G\}$$

$$G_{2,\alpha} = \{(v,\theta);\ \theta \in H_\alpha \text{ and } (v,w,\theta) \in G \text{ for some } w \in R^b\}.$$

Note that the lattices of $G_{1,\alpha}$ and $G_{2,\alpha}$ are T_α and $\pi_\alpha T$ respectively. The point groups are a subgroup $H_{1,\alpha}$ of H_α, and H_α itself.

Proposition There is a map $G_{2,\alpha} \longrightarrow (\pi_\beta T)^*/T_\beta$ with kernel $G_{1,\alpha}$. In particular $G_{1,\alpha}$ is a normal subgroup of $G_{2,\alpha}$ with abelian quotient group (cyclic if $b = 1$).

Proof Consider the assignation

$$(v,\theta) \longmapsto w \quad \text{if } (v,w,\theta) \in G.$$

Its restriction to the lattice $\pi_\alpha T$ is the map $\pi_\alpha T \to \pi_\beta T/T_\beta$ with kernel T_α discussed in the previous proposition. It has kernel $G_{1,\alpha}$ since if

(v_1,w,θ_1), (v_2,w,θ_2) are in G then $(v_1,\theta_1)(v_2,\theta_2)^{-1} \in G_{1,\alpha}$. If (v,w_1,θ), (v,w_2,θ) are in G then $w_1-w_2 \in T_\beta$ so that there is a well defined homomorphism $G_{2,\alpha} \to R^b/T_\beta$ with kernel $G_{1,\alpha}$. It remains to show that the image lies in $(\pi_\beta T)^*/T_\beta$. If $\phi \in H_\beta$ and $(x,y,\phi) \in G$ then

$$\begin{aligned}(v,w,\theta)(x,y,\phi) &= (x,y,\phi)(v,w,\theta) \bmod T\\ (v+\theta x,w+y,\theta\phi) &= (x+v,y+\phi w,\phi\theta) \bmod T\end{aligned}$$

and therefore $(x-\theta x, \phi w-w) \in T$ for all $\theta \in H_\alpha$, $\phi \in H_\beta$. It follows that $x \in (\pi_\alpha T)^*$, $w \in (\pi_\beta T)^*$.

Interchanging α and β we obtain a colour symmetry group $G_{1,\beta} \subset G_{2,\beta}$ in R^b with quotient $(\pi_\alpha T)^*/T_\alpha$. The above proof shows that the colour symmetry groups $G_{1,\alpha} \subseteq G_{2,\alpha}$ and $G_{1,\beta} \subseteq G_{2,\beta}$ must be compatible in the following sense.

<u>Definition</u> Let $G_{1,\alpha} \subseteq G_{2,\alpha}$ be a colour symmetry group in R^a with lattice $T_{1,\alpha} \subseteq T_{2,\alpha}$ and point group $H_{1,\alpha} \subseteq H_{2,\alpha}$. Let $G_{1,\beta} \subseteq G_{2,\beta}$ be a colour symmetry group in R^b. The two colour symmetry groups are *compatible* if:

(i) the groups $H_{2,\alpha}$, $H_{2,\beta}$ act trivially on $T_{2,\alpha}/T_{1,\alpha}$, $T_{2,\beta}/T_{1,\beta}$;

(ii) there are homomorphisms $G_{2,\alpha} \to T_{2,\beta}^*/T_{1,\beta}$ with kernel $G_{1,\alpha}$, and $G_{2,\beta} \to T_{2,\alpha}^*/T_{1,\alpha}$ with kernel $G_{1,\beta}$ which induce an indentification

$$T_{2,\alpha}/T_{1,\alpha} \longleftrightarrow T_{2,\beta}/T_{1,\beta}$$

and homomorphisms

$$\begin{aligned}q &: H_{2,\alpha} \longrightarrow T_{2,\beta}^*/T_{2,\beta} \quad \text{with kernel } H_{1,\alpha} \quad ,\\ q &: H_{2,\beta} \longrightarrow T_{2,\alpha}^*/T_{2,\alpha} \quad \text{with kernel } H_{1,\beta} \quad ;\end{aligned}$$

(iii) the diagram

$$\begin{array}{ccccccc} & \overset{1\times q}{\nearrow} & H_{2,\alpha} \times T_{2,\alpha}^*/T_{2,\alpha} & \xrightarrow{e} & T_{2,\alpha} & \xrightarrow{r} & T_{2,\alpha}/T_{1,\alpha}\\ H_{2,\alpha} \times H_{2,\beta} & & & & & & \updownarrow\\ & \underset{q\times 1}{\searrow} & T_{2,\beta}^*/T_{2,\beta} \times H_{2,\beta} & \xrightarrow{e} & T_{2,\beta} & \xrightarrow{r} & T_{2,\beta}/T_{1,\beta}\end{array}$$

is commutative, where e denotes evaluation maps $(\theta,x) \longrightarrow x-\theta x$ and $(w,\phi) \longrightarrow \phi w-w$ (note that these are zero for $x \in T_{2,\alpha}$ or $w \in T_{2,\beta}$), r denotes the quotient maps of the colour symmetry groups, and the vertical arrow is the identification of (ii).

Theorem Let $G_{1,\alpha} \subseteq G_{2,\alpha}$ and $G_{1,\beta} \subseteq G_{2,\beta}$ be colour symmetry groups in R^a and R^b which are compatible. The identification $T_{2,\alpha}/T_{1,\alpha} \longleftrightarrow T_{2,\beta}/T_{1,}$ determines a sublattice T of $T_{2,\alpha} \times T_{2,\beta}$ which spans $R^n = R^a \times R^b$ and which contains $T_{1,\alpha} \times T_{2,\beta}$. The homomorphisms of (ii) determine a space group G with lattice T and point group $H = H_{2,\alpha} \times H_{2,\beta}$.

Proof Let T be the set of all $(v,w) \in T_{2,\alpha} \times T_{2,\beta}$ which correspond under the identification. Then T contains $T_{1,\alpha} \times T_{1,\beta}$ and so spans R^n. Consider the group $G'_{2,\alpha}$ consisting of all (v,w,θ) for which $(v,\theta) \in G_{2,\alpha}$ maps to $w \in T^*_{2,\beta}$ modulo $T_{1,\beta}$ under the homomorphism of (ii). Then $G'_{2,\alpha}$ is isomorphic to $G_{2,\beta}$ but contains the lattice $T \subset R^n$ instead of $T_{2,\alpha} \subset R^a$. Similarly there is a group $G'_{2,\beta}$ isomorphic to $G_{2,\beta}$ and containing the lattice T. Let G be the group generated by $G_{2,\alpha}$ and $G_{2,\beta}$.

Remark If either $H_{2,\alpha}$ or $H_{2,\beta}$ is the identity then condition (iii) is automatically fulfilled because both $e(1 \times q)$ and $e(q \times 1)$ are zero. Thus suppose $H_{2,\alpha} = H$, $H_{2,\beta} = 1$. The compatibility conditions become simply

(i) H acts trivially on $T_{2,\alpha}/T_{1,\alpha}$;

(ii) there is a homomorphism $G_{2,\alpha} \to R^b/T_{1,\beta}$ with kernel $G_{1,\alpha}$ which induces an identification

$$T_{2,\alpha}/T_{1,\alpha} \longleftrightarrow T_{2,\beta}/T_{1,\beta}$$

and a homomorphism $q : H_{2,\alpha} \to R^b/T_{2,\beta}$ with kernel $H_{1,\alpha}$. Note that q may be zero even though $G_{2,\alpha} \to R^b/T_{1,\beta}$ has image larger than $T_{2,\beta}/T_{1,\beta}$.

6.3. EXAMPLES FOR n=3

The results of 6.1 and 6.2 can be used to determine a large number of 3-dimensional space groups. However care must be taken if the same space group can arise in different ways, for example from different choices of decomposition $R^a \times R^b$ or from homomorphisms q related by automorphisms of

R^a or R^b.

We first deal directly with some particularly easy arithmetic classes. If $H = 1$ or $\bar{1}$ it is clear that there is a unique isomorphism class of space groups, despite many possible choices of decomposition. If $H = \bar{3}, \bar{4}$ or $\bar{6}$ it is cyclic with a single generator $\bar{\theta}$; if $(v,\bar{\theta}) \in G$ then, by change of origin, v can be altered by

$$w - \bar{\theta}w \;=\; w + \theta w \;=\; 2w_\beta + w_\alpha + \theta w_\alpha$$

where w_β is the component of w along the axis of θ, and w_α is the component of w perpendicular to the axis, so that we may choose $v = 0$. We therefore obtain

Arithmetic crystal class	Number of space groups
$(P,1)$	1
$(P,\bar{1})$	1
$(P,\bar{3})$	1
$(R,\bar{3})$	1
$(P,\bar{4})$	1
$(I,\bar{4})$	1
$(P,\bar{6})$	1
7	7

These cases are therefore omitted in the tables which follow. The first deals with the case a=2, b=1, $H_\beta=1$. The space groups G obtained correspond to colour symmetry groups $G_{1,\alpha} \subseteq G_{2,\alpha}$ in the plane. These are listed according to the arithmetic crystal class of G for easy comparison with 5.3. The symbols θ,μ denote generating rotations, reflections for $H_{2,\alpha}$, and the symbol $\sim$ denotes equivalence resulting from the automorphism $w \mapsto \bar{w}$ of $R^b = R$. The 1-dimensional lattice $T_{1,\beta}$ is identified with the set of integers in R.

Arithmetic crystal class	$\lvert T_{2,\alpha}/T_{1,\alpha}\rvert$	$G_{2,\alpha}$	Information on $G_{1,\alpha}$, $T_{2,\beta}$ and q	Number of space groups
(P,2)	1	P2	P2 or P1 as $q(\theta) = 0$ or $\frac{1}{2}$	2
(I,2)	2	P2	P2 since $\frac{1}{2} \in T_{2,\beta} \Rightarrow q(\theta) = 0$	1
(P,3)	1	P3	P3 or P1 as $q(\theta) = 0$ or $\frac{1}{3} \sim \frac{2}{3}$	2 (+1)
(R,3)	3	P3	P3 since $\frac{1}{3} \in T_{2,\beta} \Rightarrow q(\theta) = 0$	1
(P,4)	1	P4	P4,P2,P1 as $q(\theta) = 0, \frac{1}{2}, \frac{1}{4} \sim \frac{3}{4}$	3 (+1)
(I,4)	2	P4	P4,P2 as $q(\theta) = 0, \frac{1}{4} (\frac{1}{2} \in T_{2,\beta})$	2
(P,6)	1	P6	P6,P3,P2,P1 as $q(\theta) = 0, \frac{1}{2}, \frac{1}{3} \sim \frac{2}{3}, \frac{1}{6} \sim \frac{5}{6}$	4 (+2)
(P,m)	1	Pm	Pm or P1 as $q(\mu) = 0$ or $\frac{1}{2}$	2
		Pg	equivalent to above by choice of R[b]	-
(I,m)	1	Cm	Cm or C1 as $q(\mu) = 0$ or $\frac{1}{2}$	2
(P,mm)	1	Pmm	Pmm,Pm,P2 } as $q(\mu_1)$ and $q(\mu_2)$	10
		Pgm	Pgm,Pg,Pm,P2 } are 0 or $\frac{1}{2}$;	
		Pgg	Pgg,Pg,P2 } see also 6.4	
(A,mm)	2	Pmm	Pmm or Pmg } although $\frac{1}{2} \in T_{2,\beta}$	4
		Pgm	Pgm or Pgg } and q=0; see also 6.4	
(C,mm)	1	Cmm	Cmm,Cm,C2 as $q(\mu_1), q(\mu_2)$ are 0 or $\frac{1}{2}$	3
(I,mm)	2	Cmm	Pmm,Pgm,Pgg although $\frac{1}{2} \in T_{2,\beta}$ and q=0	3
(F,mm)	2	Pmm	Cmm as $q(\mu_1), q(\mu_2)$ is 0,0 }	2
		Pgg	P2 $\frac{1}{4}, \frac{1}{4}$ }	
(P,31m)	1	P31m	P31m or P3 as $q(\mu)$ is 0 or $\frac{1}{2}$	2
(P,3m1)	1	P3m1	P3m1 or P3 as $q(\mu)$ is 0 or $\frac{1}{2}$	2
(R,3m)	3	P31m	P3m1 or P3 as $q(\mu)$ is 0 or $\frac{1}{2}$ }	2
		P3m1	P31m or P3 equivalent to above }	

(P,4m)	1	P4mm	P4mm,P2m1,P21m,P4	as $q(\mu),q(\theta)$	8
		P4gm	P4gm,Pmm,Pgg,P4	are 0 or $\frac{1}{2}$	
(I,4m)	2	P4mm	P4mm or P4gm although $\frac{1}{2}\in T_{2,\beta}$ and q=0		4
		P4gm	P4 with $q(\mu)$ = 0 or $\frac{1}{4}$		
(P,6m)	1	P6mm	P6mm,P31m,P3m1,P6 as $q(\mu),q(\theta)$ are 0 or $\frac{1}{2}$		4
20			Total number in table		63 (+4)

The proposition of 6.1 can now be used to determine further space groups. This is done in the next table. Where no explanation is given the case is covered by the more comprehensive treatment in 6.4. We use the notation of 6.1 in which K is the group generated by H and $\bar{\iota}$, and T* is the weight group of T with respect to H. It is necessary only to calculate (T*/T)/2(T*/T) and then to consult the above table to see which space groups correspond to cohomology elements of order 2.

Arithmetic crystal class	H	(T*/T)/2(T*/T) k=Z/2Z	Remarks	Number of space groups
(P,2/m)	2	k^2	all non-zero values equivalent	4
	or m	k	$H^1(H,R^3/T)$ = k	
(I,2/m)	2	k	2 × 1 = 2	2
	or m	0	$H^1(H,R^3/T)$ = k	
(P,mmm)	mm	k^2	see 6.4	16
(C,mmm)	mm	k	2 × 3 = 6	6
(I,mmm)	mm	k	see 6.4	4
(F,mmm)	mm	0	$H^1(H,R^3/T)$ = k	2
(P,$\bar{3}$1m)	3m	0	$H^1(H,R^3/T)$ = k	2
(P,$\bar{3}$m1)	3m	0	$H^1(H,R^3/T)$ = k	2

$(R,\bar{3}m)$	3m	0	$H^1(H,R^3/T) = k$	2
$(P,4/m)$	4	k	single element order 2 in $H^1(H,R^3/T)$	4
$(I,4/m)$	4	0	$H^1(H,R^3/T) = k$	2
$(P,4/mmm)$	4m	k	$H^1(H,R^3/T) = k^3$	16
$(I,4/mmm)$	4m	k	$H^1(H,R^3/T) = k$	4
$(P,6/m)$	6	0	$H^1(H,R^3/T) = k$	2
$(P,6/mmm)$	6m	0	$H^1(H,R^3/T) = k^2$	4
15	Total number in table			72

The space groups in the last table could also be obtained using the results of 6.2 with a=2, b=1, H_β=m. Although not strictly an application of 6.1 or 6.2, it is convenient to give here a direct treatment of the remaining arithmetic crystal classes (T,H) for which G(T) is decomposable. It is necessary to pay careful regard to the effect of changing the origin. We do this according to the following prescriptions:

<u>H = 222</u> Let the three rotations be ρ_1,ρ_2,ρ_3 and let the shortest non-zero vectors on rotation axes be (1,0,0),(0,1,0),(0,0,1). If $(v_i,\rho_i) \in G$ then change of origin by $(\frac{1}{2}x,\frac{1}{2}y,\frac{1}{2}z)$ changes v_1,v_2,v_3 by (0,y,z)(x,0,z),(x,y,0). Since $\rho_i^2 = \iota$ the i-th coordinate a_i of v_i may be chosen 0 or $\frac{1}{2}$. By a change of origin involving the last two coordinates we may arrange that $v_1 = (a_1,0,0)$ if T = P or $v_1 = (0,0,0)$ if T = C,I,F. Then changing the first coordinate gives $v_2 = (0,a_2,-a_3)$ if T = P, $v_2 = (0,0,-a_3)$ if T = C,I and $v_2 = (0,0,0)$ if T = F.

<u>H=32,42,62</u> Let the rotations be ρ of order 2 and θ of order q = 3,4,6 about the third coordinate axis and let the shortest non-zero vector on this axis be (0,0,1). By a change involving the first two coordinates $v_\theta = (0,0,c)$ where qc is integral if T = P, $v_\theta = (0,0,0)$ if T = R, and $v_\theta = (0,0,c)$ where c = 0 or $\frac{1}{4}$ if T = I. By a change in the third coordinate we may arrange $v_\rho = (a,b,0),(a,b,\frac{1}{2}),(a,b,\frac{1}{3})$ as convenient. Since $\theta\rho = \rho\theta^{-1}$ and $\rho^2 = \iota$ we have $v_\rho - \theta v_\rho \in T$ and $v_\rho + \rho v_\rho \in T$; the first equation shows that a change of origin by v_ρ will not affect v_θ, though it will replace v_ρ by $3v_\rho$ and hence $v_\rho = (0,0,0)$ for q = 3,6.

$H = \bar{4}2, \bar{6}2$ Let the generators be ρ and $\bar{\theta}$. By the argument used at the beginning of this section we may assume $v_{\bar{\theta}} = (0,0,0)$, $v_\rho = (a,b,c)$. We have $v_\rho + \theta v_\rho \in T$, $v_\rho + \rho v_\rho \in T$.

Arithmetic crystal class	Possible values for a,b,c	Number of space groups
(P,222)	a_1, a_2, a_3 interchangeable $(\frac{1}{2},\frac{1}{2},\frac{1}{2}),(\frac{1}{2},\frac{1}{2},0)\,(\frac{1}{2},0,0),(0,0,0)$	4
(C,222)	$a_3 = 0,\frac{1}{2}$	2
(I,222)	$a_3 = 0,\frac{1}{2}$	2
(F,222)	$a_3 = 0$	1
(P,312)	$a,b = 0;\ c = 0,\ \frac{1}{3}\sim\frac{2}{3}$	2(+1)
(P,321)	$a,b = 0;\ c = 0,\ \frac{1}{3}\sim\frac{2}{3}$	2(+1)
(R,32)	$a,b,c = 0$	1
(P,42)	$(a,b) = (0,0),(\frac{1}{2},\frac{1}{2});\ c = 0,\frac{1}{2},\frac{1}{4}\sim\frac{3}{4}$	6(+2)
(I,42)	$(a,b) = (0,0);\ c = 0,\frac{1}{4}$	2
(P,62)	$(a,b) = (0,0);\ c = 0,\frac{1}{2},\frac{1}{3}\sim\frac{2}{3},\ \frac{1}{6}\sim\frac{5}{6}$	4(+2)
(P,$\bar{4}$m2)	$(a,b) = (0,0),(\frac{1}{2},\frac{1}{2});\ c = 0,\frac{1}{2}$	4
(I,$\bar{4}$m2)	$(a,b,c) = (0,0,0)$ or $(\frac{1}{2},0,\frac{1}{4})$	2
(P,$\bar{4}$2m)	$(a,b) = (0,0),(\frac{1}{2},\frac{1}{2});\ c = 0,\frac{1}{2}$	4
(I,$\bar{4}$2m)	$(a,b,c) = (0,0,0)$ or $(0,0,\frac{1}{2})$	2
(P,$\bar{6}$m2)	$a = b = 0;\ c = 0,\frac{1}{2}$	2
(P,$\bar{6}$2m)	$a = b = 0;\ c = 0,\frac{1}{2}$	2
16	Total number in table	42(+6)

The above totals cover 7 + 63 + 72 + 42 = 184 space groups belonging to 7 + 20 + 15 + 16 = 58 arithmetic crystal classes (in both cases about four-fifths of the final total). The numbers in brackets refer to additional equivalence classes obtained if identifications under reflection $w \mapsto \bar{w}$ of R^b are not used; this is equivalent to saying that G is equivalent to G' if $G' = fGf^{-1}$ for some *orientation preserving* affine transformation f. This is the equivalence relation customarily used by physicists and chemists; it leads to a final total of 230 equivalence classes. The remaining arithmetic crystal classes all have cubic Bravais type; they are discussed in 6.6.

6.4. ORTHOGONAL SPACE GROUPS

The above tables included little or no discussion in the case of point groups mmm and mm. This is because they can be dealt with easily as special cases of an algorithm which is valid for arbitrary dimensions n and which takes account of the fact that the normaliser of the point group consists of coordinate permutations.

Let μ_i denote reflection in the hyperplane perpendicular to the i-th axis, and let $2e_i$ be a shortest non-zero vector of T on the i-th axis. We shall consider the cases

H generated by $\mu_1,\ldots,\mu_n$: the *holohedral case* with $|H| = 2^n$

H generated by $\mu_1,\ldots,\mu_{n-1}$: the *hemihedral case* with $|H| = 2^{n-1}$,

in this section. Recall from 5.5 that the lattice T is determined by an equivalence class of subspaces $W \subset k^n$ containing no basis vectors, where $k = \{0,1\}$. The appropriate equivalence relation giving the pair (T,H) is $W \sim W'$ if there is $\sigma \in S_n$ (holohedral case) or $\sigma \in S_n$ (hemihedral case) acting on k^n by permutation of coordinate axes such that $W' = \sigma W$. We denote by $E_{j_1 \ldots j_p}$ the subspace of k^n consisting of all vectors with an even number of coordinates $j_1,\ldots,j_p$ equal to 1 and all other coordinates 0; it has dimension p-1.

Proposition The origin can be chosen so that $(v_i,\mu_i) \in G$ where the i-th coordinate of v_i is zero.

Proof If not, replace G by $(x,\iota)G(x,\iota)^{-1}$ which contains $(x-\mu_i x+v_i,\mu_i)$. The result follows because $x - \mu_i x$ lies on the i-th coordinate axis.

Assume the origin chosen in this way. Then there are elements r_{ij}, $s_{ij} \in k = \{0,1\}$ such that

$$v_i = \sum_{j \neq i} \tfrac{1}{2}(r_{ij} + 2s_{ij})e_j.$$

Here $(r_{i1}\ldots 0\ldots r_{in}) \in W_1$ and $(s_{i1}\ldots 0\ldots s_{in})$ is determined modulo the projection $\pi_i W$ of W to the i-th coordinate hyperplane. The relation $\mu_i\mu_j = \mu_j\mu_i$ implies

$$v_i + \mu_i v_j = v_j + \mu_j v_i \text{ modulo } T$$

and therefore

$$r_{ij}e_j - r_{ji}e_i \in T.$$

Thus $r_{ij} = 1$ implies $r_{ji} = 1$. Suppose that, for fixed i, the coefficients $r_{ij_1},\ldots,r_{ij_p}$ are equal to 1 and the remaining r_{ij} are 0. Then $E_{ij_1\ldots j_p} \subseteq W$ and p is even (otherwise it would follow that W contains the i-th basis vector.

In order to take account of the action of S_n on coordinate axes, we represent the elements r_{ij}, s_{ij} of k by means of a graph with n vertices 1,...,n.

The graph is constructed by the prescription: if $r_{ij} = r_{ji} = 1$ then joing vertex i to vertex j by an undirected edge; if $s_{ij} = 1$ then join vertex i to vertex j by a directed edge.

Because the vector $(s_{i1}\ldots 0\ldots s_{in})$ is determined only modulo $\pi_i W$, we apply the corresponding equivalence relation to the graphs: the set of directed edges leaving a given vertex i may be modified by any $w \in \pi_i W$ (that is, if w has non-zero coordinates $j_1,\ldots,j_q$ then the subset of edges ij_α is replaced by the complementary subset of edges ij_β in the set $\{ij_1,\ldots ij_q\}$). It is a consequence that if $E_{j_1\ldots j_p} \subseteq W$ then there is an equivalent graph in which no directed edges run between vertices $j_1,\ldots,j_p$ and in which any edge arriving at a vertex j_α from outside the set may be replaced by an edge arriving at any other j_β.

It follows that the prescription gives a well defined equivalence class of graphs with no ambiguities arising from undirected and directed edges. The former constitute an *even* graph whose edges are restricted to pairs j_α, j_β where $E_{j_1 \dots j_p} \subseteq W$. The latter are subject to an equivalence relation if $W \neq \{0\}$ but never include an edge of the form $j_\alpha j_\beta$. In the hemihedral case one vertex is distinguished and no directed edges depart from it.

Theorem There is a one-one correspondence between isomorphism classes of n-dimensional orthogonal space groups G and equivalence classes of

(i) graphs Γ with n vertices, in the holohedral case,

(ii) graphs Γ with n vertices, one of which is distinguished and from which no directed edges depart, in the hemihedral case.

The graphs and equivalence relation are of the following kind. There are distinguished sets of vertices $\{j_1, \dots, j_p\}$ corresponding to summands $E_{j_1 \dots j_p} \subseteq W$; for each such set Γ contains an even subgraph of undirected edges between vertices j_α, j_β of the set. The remaining edges of Γ are directed and never include any of the form $j_\alpha j_\beta$. The directed edges departing from a given vertex i may be replaced by an equivalent set of edges also departing from i but with arrivals differing by an element $w \in \pi_i W$.

Remarks (i) Orthogonal space groups with a choice of *setting*, or coordinate basis $e_1, \dots, e_n$ correspond to equivalence classes of graphs with labelled vertices. (ii) The content of the theorem is best made clear by examples; in the diagrams below the distinguished sets are circled, while in the hemihedral case the distinguished vertex is $\circ$.

Examples (n=2). There are 4 holohedral and 3 hemihedral graphs:

Holohedral case

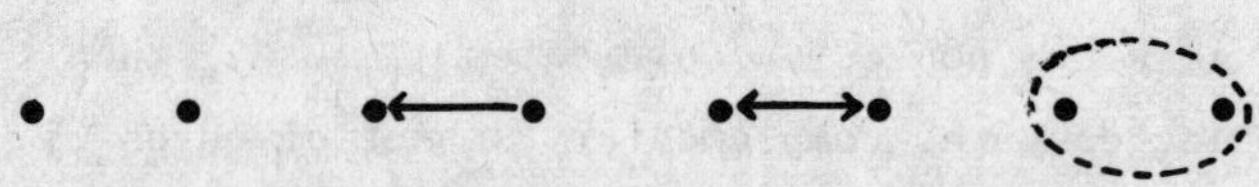

Pmm Pmg Pgg Cmm

Hemihedral case

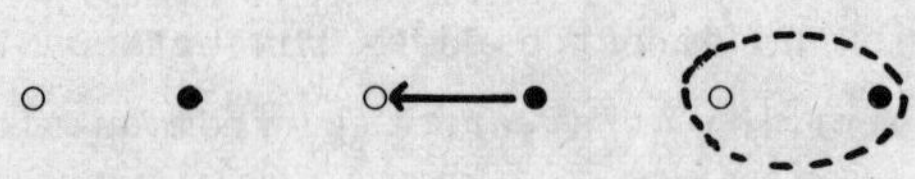

Pm Pg Cm

Examples (n=3) There are 28 holohedral and 22 hemihedral graphs:

Holohedral case (primitive lattice)

16 graphs of class (P,mmm)

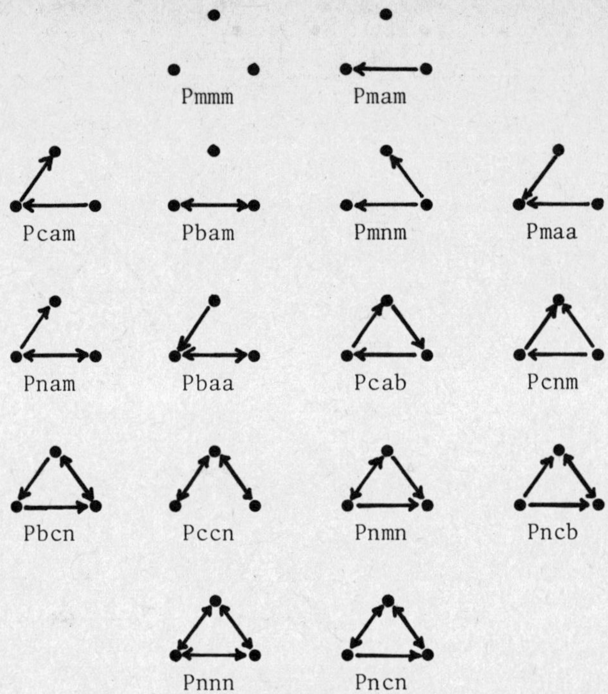

[Note that the graph is determined uniquely by the space group, whereas the corresponding name, e.g. Pcam, is not determined uniquely. There can be up to n! names corresponding to different ways of naming the vertices. The above names correspond to the setting c● a● ●b]

Holohedral cases (other lattices)

6 graphs of class (C,mmm)

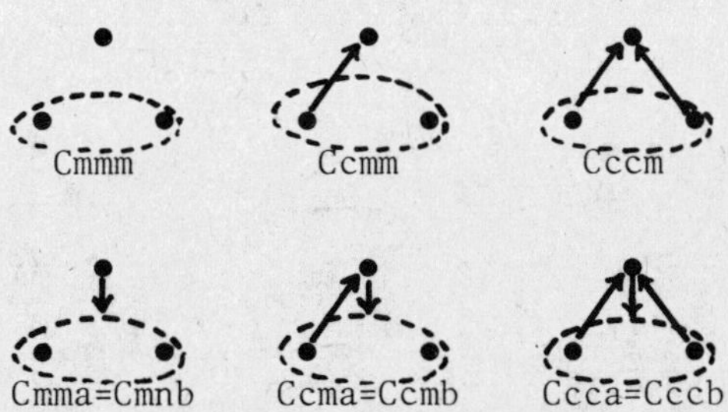

4 graphs of class (I,mmm)

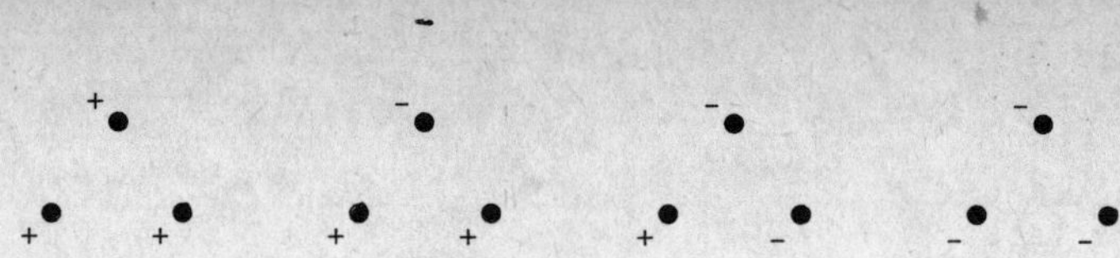

Immm etc   Imma etc   Imaa etc   Ibca etc

[the symbol + or - indicates whether even or odd number of edges leave the vertex]

2 graphs of class (F,mmm)

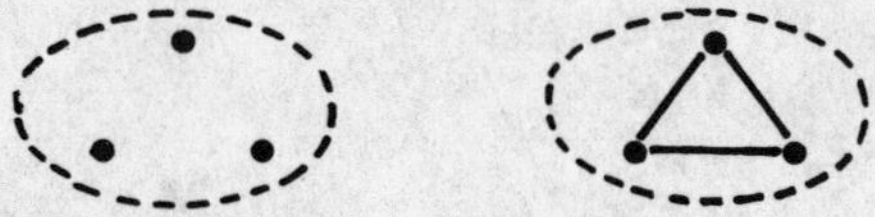

Fmmm   Fddd

Hemihedral case (primitive lattice)

Pmm   Pma   Pmc

10 graphs of class (P,mm)

Pba   Pnm   Pbc   Pcc

Pnn   Pbn   Pnc

Hemihedral case (other lattices)

3 graphs of class (C,mm)

Cmm   Ccm   Ccc

4 graphs of class (A,mm)

Amm   Acm   Ama =Amb   Aca =Acb

3 graphs of class (I,mm)

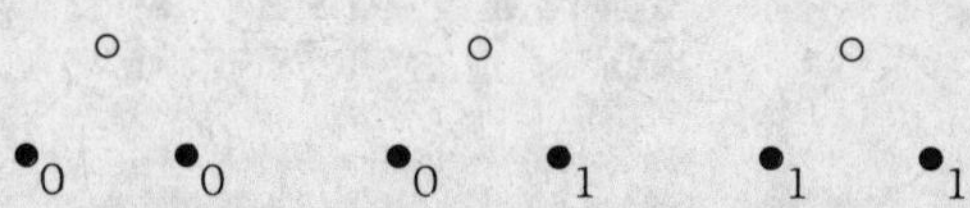

Imm etc   Ima etc   Iba etc

2 graphs of class (F,mm)

Examples (n=4) There are 419 holohedral and 294 hemihedral graphs, and therefore not all will be drawn, but we indicate how the various graphs arise in the holohedral case (for details of the 8 lattices see 5.5):

218 graphs of class (P,mmmm) corresponding to the 218 directed graphs with 4 vertices.

88 graphs of class (S,mmmm) as indicated; only edges leaving from the circled set have been drawn.

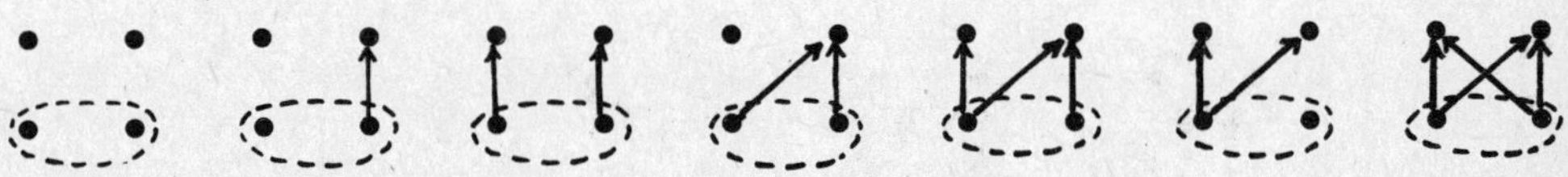

10 graphs, 16 graphs, 10 graphs, 16 graphs, 16 graphs, 10 graphs, 10 graphs

60 graphs of class (I,mmmm) as indicated; the equivalence relation means that we may assume that 0 or 1 edge leaves the vertex not involved in the I-centring, only edges involving this vertex have been drawn.

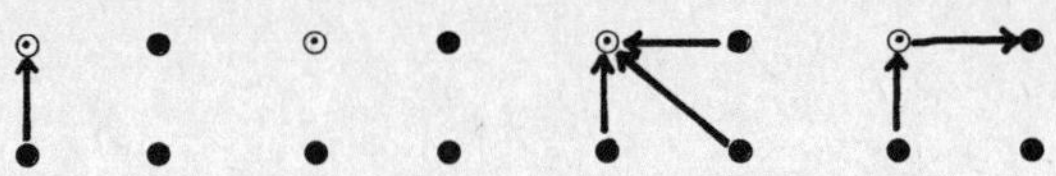

6 graphs, 4 graphs, 4 graphs, 6 graphs

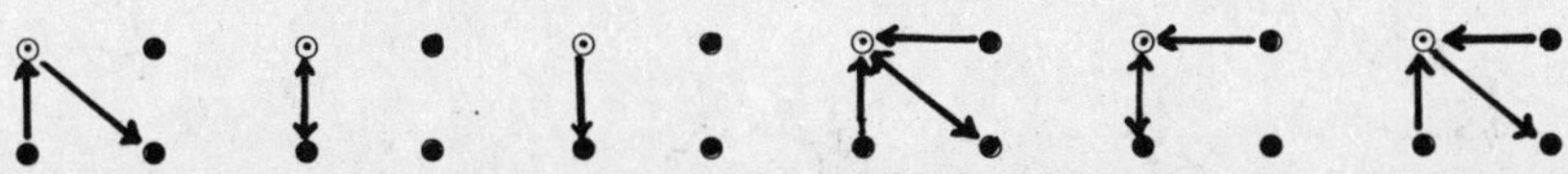

8 graphs, 6 graphs, 6 graphs, 6 graphs, 8 graphs, 6 graphs

19 graphs of class (Z,mmmm) corresponding to the 19 directed graphs with 4 vertices for which the number of edges leaving each vertex is even (which can always be arranged by virtue of the equivalence relation).

16 graphs of class (F,mmmm) as indicated; edges leaving vertices in the circled set have not been drawn.

4 graphs 4 graphs 4 graphs 4 graphs

6 graphs of class (D,mmmm) as indicated

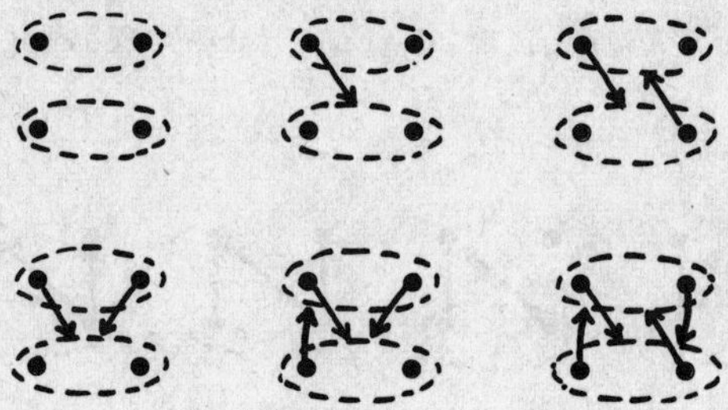

9 graphs of class (G,mmmm) as indicated; the equivalence relation can be used to identify the two possible edges leaving any vertex.

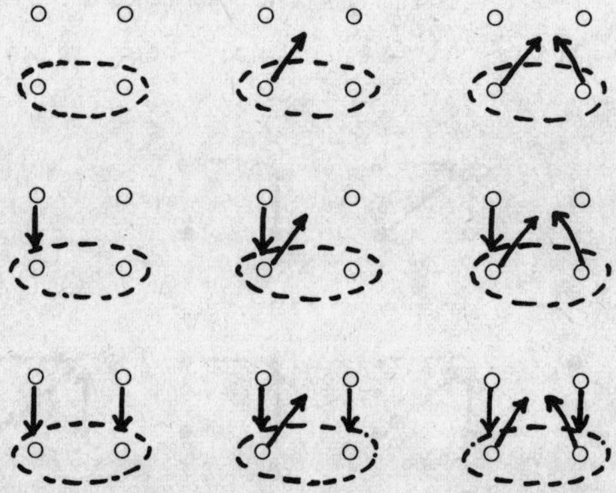

3 graphs of class (U,mmmm) as indicated

The number of directed graphs in other low dimensions is known, and can be computed for any desired value n by use of Polya's enumeration theorem. Some resulting numbers of orthogonal space groups for T primitive are:

n	primitive holohedral	primitive hemihedral
5	9608	3044
6	1540 944	291 968
7	882 033 440	96 928 992
asymptotic value for large n	$2^{n(n-1)}/n!$	$2^{(n-1)^2}/(n-1)!$

Let T be a lattice corresponding to a r-dimensional subspace $W \subset k^n$. There is a (possibly empty) set of summands of the form $E_{i_1 \dots i_\ell}$ in W; these determine a list Λ of integers ℓ with $\sum_{\ell \in \Lambda} (\ell-1) \le r$. The set of *labelled* graphs (i.e. graphs before action by permutations of vertices) can be regarded as a vector space over k of dimension

$$n(n-1-r) + \tfrac{1}{2} \sum_{\ell \in \Lambda} (\ell-1)(\ell-2) \quad \text{in the holohedral case,}$$

$$(n-1)(n-1-r) + \tfrac{1}{2} \sum_{\ell \in \Lambda} (\ell-1)(\ell-2) \quad \text{in the hemihedral case.}$$

Since there is no contribution to the dimension from $\ell = 1,2$ we shall assume $\ell > 2$ for $\ell \in \Lambda$; this implies $\sum_{\ell \in \Lambda} (\ell-1)(\ell-2) \le r(r-1)$. Let $d_{r,n}$ be the number of distinct r-dimensional subspaces which contain no basis vector. Since $d_{o,n} = 1$ and $d_{1,n} = 2^n-n-1$, the total number of labelled graphs which can occur as graphs of orthogonal space groups with dim W < 2 is

$$2^{n(n-1)}+2^{n(n-2)}(2^n-n-1) = 2^{n(n-1)}(2-\frac{n+1}{2^n}) \quad \text{in the holohedral case,}$$

$$2^{(n-1)(n-1)}+2^{(n-1)(n-2)}(2^n-n-1) = 2^{n(n-2)}(6-\frac{n+1}{2^{n-2}}) \quad \text{in the hemihedral case.}$$

For dim W = r > 2 the number of labelled graphs is bounded above by $d_{r,n}2^{n(n-2)-rn+\frac{1}{2}r(r-1)+r+1}$ in the holohedral case, and by

$d_{r,n}2^{n(n-2)-rn+\frac{1}{2}r(r-1)+r+1}$ in the hemihedral case. The inequality

$$d_{r,n} < \frac{(2^n-1)(2^{n-1}-1)\ldots(2^{n-r+1}-1)}{(2^r-1)(2^{r-1}-1)\ldots(2-1)} < \frac{2^{nr-\frac{1}{2}r(r-1)}}{(2^r-1)(2^{r-1}-1)\ldots(2-1}$$

leads to upper bounds

$$2^{n(n-1)} \sum_{r=2}^{n} \frac{1}{(2^r-1)\ldots(2-1)} < 2^{n(n-1)} (\tfrac{2}{5}) \text{ in the holohedral case,}$$

$$2^{n(n-2)} \sum_{r=2}^{n} \frac{2^{r+1}}{(2^r-1)\ldots(2-1)} < 2^{n(n-2)} (4) \text{ in the hemihedral case.}$$

To obtain asymptotic values for the number of orthogonal space groups it is necessary to consider the action of permutations on labelled graphs. For $W = \{0\}$ this gives approximations $2^{n(n-1)}/n!$ and $2^{n(n-2)+1}/(n-1)!$ for the primitive holohedral and hemihedral cases. For dim $W > 0$ the corresponding expressions may be too small, but in any case we have

<u>Theorem</u> Let s_n be the logarithm to base 2 of the number of isomorphism classes of orthogonal n-dimensional space groups. Then $\lim_{n\to\infty} (s_n/n^2) = 1$.

6.5. ORTHOGONAL SPACE GROUPS WITH ADDITIONAL SYMMETRY

Let $A \subseteq S_n$ act by permutation of coordinates on k^n. It permutes coordinate axes of R^n. We consider space groups for which the point group is generated by the reflections $\mu_1,\ldots,\mu_n$ (holohedral case) or $\mu_1,\ldots,\mu_{n-1}$ (hemihedral case) together with coordinate permutations of A. If it is necessary to consider actions on R^n we choose a covering $\tilde{A} \to A$ which may be of higher order so that $\tilde{\alpha} \in \tilde{A}$ is orthogonal. If $m \neq \ell$ and $\alpha \in A$ then $\tilde{\alpha}\mu_\ell\tilde{\alpha}^{-1}$ maps

$$e_{\alpha(m)} \longmapsto \pm e_m \longmapsto \pm e_m \longmapsto e_{\alpha(m)}$$

$$e_{\alpha(\ell)} \longmapsto \pm e_\ell \longmapsto \mp e_\ell \longmapsto -e_{\alpha(\ell)}$$

and therefore $\tilde{\alpha}\mu_\ell\tilde{\alpha}^{-1} = \mu_{\alpha(\ell)}$. It follows that in the hemihedral case

$A \subseteq S_{n-1}$ leaves the n-th coordinate axis invariant. In both cases $(w,\tilde{\alpha}) \in G$ implies

$$(w,\tilde{\alpha})(v_i,\mu_i)(w,\tilde{\alpha})^{-1} = (\tilde{\alpha}v_i+w-\mu_{\alpha(i)}w,\ \mu_{\alpha(i)})$$

is in G and therefore

$$w-\mu_{\alpha(i)}w = v_{\alpha(i)} - \tilde{\alpha}v_i \quad \text{modulo T.}$$

But the right hand side of the equation has $\alpha(i)$-th coordinate zero, and has an even number of coordinates congruent to $\frac{1}{2}$ modulo 1. Therefore the left hand side is 0 modulo $e_{\alpha(i)}$ and the $\alpha(i)$-th coefficient of w is 0 or $\frac{1}{2}$ modulo 1. In the notation of 6.4

$$(r_{\alpha(i)1}\ldots\ldots r_{\alpha(i)n}) = \alpha(r_{i1}\ldots\ldots r_{in}),$$

$$(s_{\alpha(i)1}\ldots\ldots s_{\alpha(i)n}) \equiv \alpha(s_{i1}\ldots\ldots s_{in}) \text{ modulo } \pi_{\alpha(i)}W,$$

where now A is regarded as acting by permutation of axes on k^n.
We have

<u>Proposition</u> Let G be an orthogonal space group with additional symmetry A, and let Γ be the graph with n vertices determined by G. Then the graphs Γ, $\alpha\Gamma$ are equivalent (and hence equal if $W = \{0\}$).

Of course the equivalence class of graphs does not now determine the space group G; it is necessary also to know the vectors w_α such that $(w_\alpha,\tilde{\alpha}) \in G$. In many cases, however, this information can be displayed by colouring of the graph Γ, and examples of this are given below.

The action of A on k^n by permutation of coordinate axes does not determine the action of $\tilde{A}$ on R^n since $\tilde{\alpha}e_i$ may be $+e_{\alpha(i)}$ or $-e_{\alpha(i)}$. Except in one situation, it is always possible, by appropriate composition with reflections μ_i, to choose generators $\tilde{\alpha}$ such that $\tilde{\alpha}e_i$ has a desired sign. The exception occurs in the hemihedral case when i=n is the distinguished vertex. We write $R^n = R^{n-1} \times R^1$ and recall that a shift of origin by $(b,c) \in R^{n-1} \times R^1$ will leave the vectors v_i unaltered provided that $2b \in T$.

There are two distinct kinds of action of $\tilde{A}$ on $R^{n-1} \times R^1$ in the hemihedral case. (i) $\tilde{\alpha} e_n = +e_n$ for all $\alpha \in A$. The vectors w_α such that $(w_\alpha, \tilde{\alpha}) \in G$ may be altered by $(b-\alpha b, 0)$ provided that $2b \in T$. Here b is fixed, while $\tilde{\alpha}$ varies over all $\alpha \in A$.
(ii) $\tilde{\alpha} e_n = +e_n$ for $\alpha \in A_o \subset A$, and $\tilde{\alpha} e_n = -e_n$ for $\alpha \in A \backslash A_o$. Now $\{w_\alpha\}_{\alpha \in A}$ may be altered by $\{(b-\tilde{\alpha}b, 0);\ \alpha \in A_o\}$ and $\{(b-\tilde{\alpha}b, 2c);\ \alpha \in A\backslash A_o\}$ provided that $2b \in T$; in particular it can be arranged that $w_\alpha = (y, 0)$ for some definite choice of generator $\alpha \in A\backslash A_o$.

Similarly in the holohedral case $\{w_\alpha\}_{\alpha \in A}$ may be altered by $\{u-\tilde{\alpha}u\}$ for some fixed u such that $2u \in T$.

We illustrate the above situation by enumerating various tetragonal orthogonal groups.

Example Let A be generated by $\alpha = (12)$, and define $\tilde{\alpha}$ by $\tilde{\alpha} e_1 = e_2$, $\tilde{\alpha} e_2 = e_1$, $\tilde{\alpha} e_i = e_i$ for $i > 2$. Let $w = (x,y,z) \in R \times R \times R^{n-2} = R^n$. Then

$$w + \tilde{\alpha} w = (x+y, x+y, 2z) \in T$$

Primitive tetragonal groups In this case $W = \{0\}$ so $v_1 = \tilde{\alpha} v_2$ and x,y must take values 0 or 1. By the above equation the only possibilities are (0,0) or (1,1). But $u = (0,1)$ gives $(1,1) = u - \tilde{\alpha} u$ modulo T. Therefore we may assume $w = (0,0,z)$ where $z \in R^{n-2}$ satisfies $2z \in T$. Values of z may be displayed by colouring the i-th vertex whenever the i-th coordinate of z is 1 for $i = 3,4,\ldots,n$. For n=3 there are 16 possibilities in the holohedral case; they correspond to the 8 graphs below invariant under action of A with vertex 3 (marked ○) either uncoloured or coloured. Vertices 1,2 are marked ⊙.

(z=0) P4/mmm P4/nmm P4/mcm P4/ncm

(z=0) P4/mbm P4/nbm P4/mnm P4/nnm

(z=1) P4/mbc P4/nbc P4/mnc P4/nnc

In the hemihedral case there are 12 possibilities only, since now no edges may depart from vertex 3 and z=0 if $\tilde{\alpha}e_3 = -e_3$.

	$\tilde{\alpha}e_3 = +e_3$		$\tilde{\alpha}e_3 = -e_3$	
	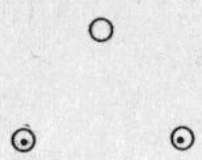		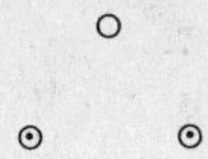	
(z=0)	P4mm	P4cm	P$\bar{4}$m2	P$\bar{4}$c2
(z=1)	P4mc	P4cc		
	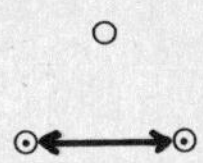	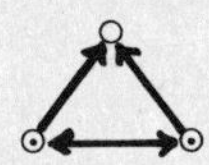	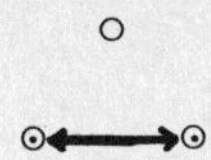	
(z=0)	P4bm	P4nm	P$\bar{4}$b2	P$\bar{4}$n2
(z=1)	P4bc	P4nc		

For n=4 there are 272 possibilities in the holohedral case; they correspond to 72 graphs with vertices 3,4 (marked ●) either uncoloured or coloured. Vertices 1,2 are marked ⊙ and only edges between ⊙ and ○ have been drawn.

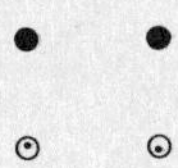   

each × 2 ways of connecting ⊙ × 3 ways of connecting ●, giving 4 × 2 × 3 = 24 distinct graphs of which 16 are invariant under interchange of vertices ● and 8 are not.

 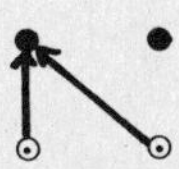 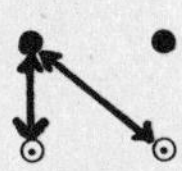   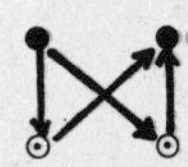

each × 2 ways of connecting ⊙ × 4 ways of connecting ●, giving 6 × 2 × 4 = 48 distinct graphs all not invariant under interchange of vertices ●. The total number of ways of colouring vertices ● is (16 × 3)+(8 + 48) × 4 = 272. In the hemihedral case there are 192 possibilities corresponding to the graphs

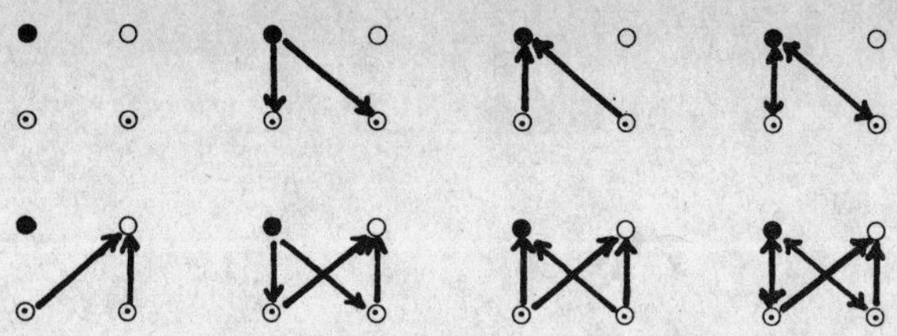

each × 2 ways of connecting ⊙ × 2 ways of connecting ● to ○. (32 graphs in all). If $\tilde{\alpha}e_4 = +e_4$ there are 4 ways of colouring the vertices 3,4 giving 32 x 4 = 128 space groups. If $\tilde{\alpha}e_4 = -4$ there are 2 ways of colouring the vertex 3 giving 32 x 2 = 64 space groups.

The primitive tetragonal groups are of interest because in the holohedral case they form the largest arithmetic crystal class for certain exceptional values of n (although for large n the class is smaller than that of primitive orthogonal groups). Comparative statistics of isomorphism classes of space groups in these dimensions are

	primitive orthogonal		primitive tetragonal orthogonal		
n	holohedral	hemihedral	holohedral	hemihedral $\tilde{\alpha}e_n = +e_n$	hemihedral $\tilde{\alpha}e_n = -e_n$
3	16	10	16	8	4
4	218	104	272	128	64
5	9608	3044	11456	4224	2112
lower bound and asymptotic value	$\dfrac{2^{n(n-1)}}{n!}$	$\dfrac{2^{n(n-2)+1}}{(n-1)!}$	$\dfrac{2^{n(n-2)+1}}{(n-2)!}$	$\dfrac{2^{n(n-3)+3}}{(n-3)!}$	$\dfrac{2^{n(n-3)+2}}{(n-3)!}$

In all cases convergence to the asymptotic value is very rapid.

<u>Body centred tetragonal groups</u> We suppose n=3 and W = {(000)(111)}. In the holohedral case w must have $\alpha(i)$-th coordinate 0 if $v_{\alpha(i)} = \tilde{\alpha}v_{(i)}$ and $\frac{1}{2}$ if $v_{\alpha(i)} \neq \tilde{\alpha}v_i$, and so is completely determined by the graph.

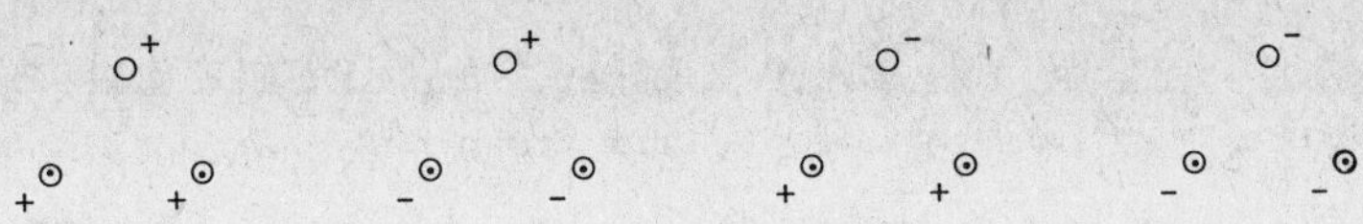

I4/mmm I4/mcm I4/amd I4/acd

In the hemihedral case there are 6 space groups (corresponding to graphs with colouring of vertex 3 if $\tilde{\alpha}e_3 = e_3$).

$\tilde{\alpha}e_3 = +e_3$ $\tilde{\alpha}e_3 = -e_3$

(z=0)	I4mm	I4cm	I$\bar{4}$m2	I$\bar{4}$c2
(z=½)	I4md	I4cd		

6.6. COHOMOLOGY OF PERMUTATION GROUPS

The determination of tetragonal orthogonal space groups in 6.5 is an example of a general procedure for the calculation of the possible vectors w_α such that $(w_\alpha, \tilde{\alpha}) \in G$ for $\alpha \in A \subseteq S_n$ acting on coordinate axes. We assume $\tilde{\alpha}$ chosen so that $\tilde{\alpha}e_i = e_i$ for all i except possibly i=n in the hemihedral case; this can be arranged by composition if necessary with reflections and implies that $\tilde{A}$ may be identified with A.

Definition Let $A \subseteq S_n$ act on k^n by permutation of coordinates and let $W \subset k^n$ be a subspace invariant under A. An operator homomorphism $w : A \to k^n/W$ is a map satisfying $w(\alpha\beta) = w(\alpha) + \alpha w(\beta)$. It is trivial if there exists $x \in k^n/W$ such that $w(\alpha) = x - \alpha x$ for all $\alpha \in A$. The additive group of all operator homomorphisms modulo those which are trivial is the (first) cohomology group $H^1(A, k^n/W)$.

Proposition Consider an A-invariant subspace $W \subset k^n$ and an equivalence class of graphs containing some Γ with $v_{\alpha(i)} - \alpha v_i \in W$ for all $\alpha \in A$ and $i = 1,\ldots,n$. The corresponding isomorphism classes of space groups in the holohedral case are in one-one correspondence with the elements of $H^1(A, k^n/W)$.

Proof By 6.4 the vectors w_α such that $(w_\alpha, \alpha) \in G$ satisfy $w_\alpha - \mu_{\alpha(i)} w_\alpha \in T$ for all $\alpha \in A$ and $i = 1,\ldots,n$. Therefore every coefficient of w_α is 0 or 1 modulo 2, and w can be regarded as a map $A \to k^n/W$ satisfying $w(\alpha\beta) = w(\alpha) + \alpha w(\beta)$. Two space groups with the same values v_i are isomorphic if and only if they are conjugate with respect to shift of origin by $u \in R^n$ with $u - \mu_i u \in T$ for $i = 1,\ldots,n$. Therefore every coefficient of u is 0 or 1 modulo 2 and u corresponds to $x \in k^n/W$ changing $w(\alpha)$ by $x - \alpha x$.

Remarks (i) The assumption about Γ always holds in the primitive case $W = \{0\}$, and more generally whenever $\Gamma = \alpha\Gamma$ since then $v_{\alpha(i)} - \alpha v_i = 0$. When n=3 and W = (000)(111), corresponding to the body centred lattice I, the assumption holds for some equivalence classes and not for others

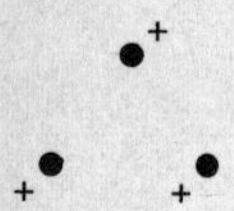 holds for all A by choosing Γ = 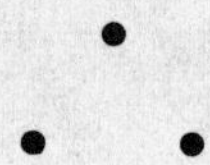 or 

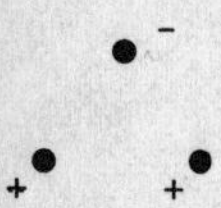 does not hold for A = <(12)>

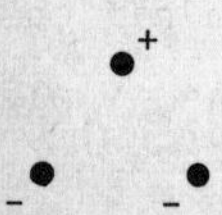 holds for A = <(12)> by choosing 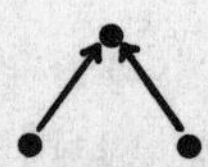 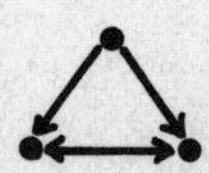

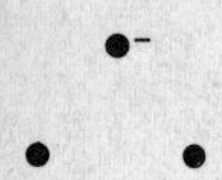 holds for A = <(123)> by choosing 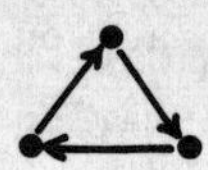, not for $A = S_3$.

(ii) For particular A,W for which the assumption does not hold similar results may well still hold. For example if n=3 and w = (000)(111) then if 3 is a vertex marked '-', that is with one edge departing, it follows that the transposition $\alpha = (12)$ gives $v_{\alpha(i)} - \alpha v_i = e_3$ modulo T and so w has 3rd coordinate $\frac{1}{2}$ modulo 1. Any two choices for w differ by an element of $H^1(A,k^3/W)$.

(iii) In the hemihedral case the assumption becomes $v_{\alpha(i)} - \alpha v_i \in W$ for $i = 1,\ldots,n-1$; the first n-1 coordinates of w determine an element of

$H^1(A, k^{n-1}/\pi_n W)$, while the last coordinate determines a homomorphism $A_o \to R/Z$. Here $A_o \subseteq A$ is the subgroup of α with $\alpha e_n = +e_n$, and $Z \subset R$ is identified with the set of multiples of e_n, if $(r_1 \ldots r_n 1) \in W$, and of $2e_n$ otherwise.

The calculation of $H^1(A,k^n/W)$ can be simplified by use of the divisibility theorem (see 2.4 and Appendix). If $A_1 \subseteq A$ is a subgroup of odd order then every element of $H^1(A,k^n/W)$ may be represented by an operator homomorphism w which is zero on A_1. Explicitly, if w' is another choice let $w(\alpha) = w'(\alpha) - x + \alpha x$ where

$$x = \frac{1}{|A_1|} \sum_{\alpha_1 \in A_1} w'(\alpha_1).$$

Assume that this has been done and that A is generated by $\alpha, \beta, \ldots$ over A_1. It is now sufficient to determine $w(\alpha), w(\beta), \ldots \in k^n/W$. Each such element can be regarded as a class modulo W of colourings of vertices of Γ : each vertex corresponding to a non-zero coordinate of $w(\alpha)$ is marked with colour 'α'.

It is necessary to ensure that the colouring of vertices is consistent with relations between generators. Such a relation can be written

$$\alpha_1 \alpha_2 \ldots . \alpha_k = \iota$$

where each α_i is a generator α, or the inverse α^{-1} of a generator (in which case set $w(\alpha_i) = \alpha^{-1} w(\alpha)$), or in A_1 (in which case $w(\alpha_i) = 0$). For each vertex v the relation determines a circuit $u_k = \alpha_k v$, $u_{k-1} = \alpha_{k-1}\alpha_k v, \ldots$, $u_2 = \alpha_2 \ldots \alpha_{k-1}\alpha_k v$, $u_1 = v$, which is represented diagramatically below; in fact some of the edges may be loops and some of the vertices may coincide.

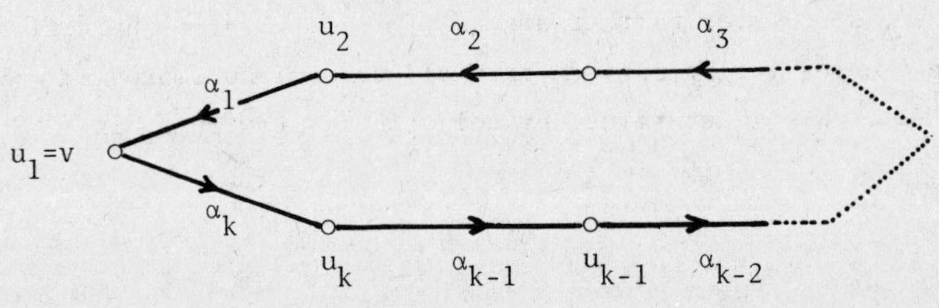

Definition The vertex u_i is *distinguished* (for the circuit) if it is marked with colour 'α_i', that is if the u_i-th coordinate of $w(\alpha_i)$ is non-zero.

Proposition Suppose that W contains no vector with an odd number of coordinates $u_1,\ldots,u_k$ non-zero. Then the number of distinguished vertices in each circuit is even.

Proof. The vertex u_i is distinguished if and only if $\alpha_1\ldots\alpha_{i-1}w(\alpha_i)$ has non-zero coordinate v. But the sum of all such terms for $i = 1,\ldots,k$ is $w(\alpha_1\ldots\alpha_k) = w(\iota) = 0$ modulo W.

Corollary Let α be of order q. Then the same assumption on W implies that the number of vertices $v, \alpha v,\ldots,\alpha^{q-1}v$ which are marked with the colour 'α' is even.

Remarks (i) The corollary is true in some situations that the assumption on W fails. For example if W contains a vector w_o with non-zero coordinates $v, v_1,\ldots,v_\ell$ and $\alpha v_i = v_i$ for $i = 1,\ldots,\ell$ then $w(\alpha)$ may be altered by w_o. This changes the parity of non-zero coordinates in the set $v,..,\alpha^{q-1}v$ but not in any other α-circuit. It can therefore be arranged that the number of non-zero coordinates is even.

(ii) Note that a circuit may contain repetitions of isolated vertices, or of all vertices and edges. If the number of repetitions of the whole circuit is even then the proposition gives no information. But if the number is odd it can often be used to prove that an operator homomorphism must be trivial.

Examples (i) The tetragonal case treated in 6.4. There is one α-circuit with 2 vertices, so both or neither are non-zero for $w(\alpha)$. The difference between these two possibilities is a trivial operator homomorphism. So the cohomology $H^1(A,k^n/W)$ is determined by colouring of the remaining n-2 vertices.

α-circuits trivial operator homomorphisms

(ii) The remaining subgroups $A \subseteq S_3$ give space groups belonging to certain cubic arithmetic crystal classes not considered in 6.3. The subgroup A_3 has odd order and therefore $H^1(A_3, k^3/W) = 0$ for all W. There is therefore only one isomorphism class of space groups for each choice of graph Γ

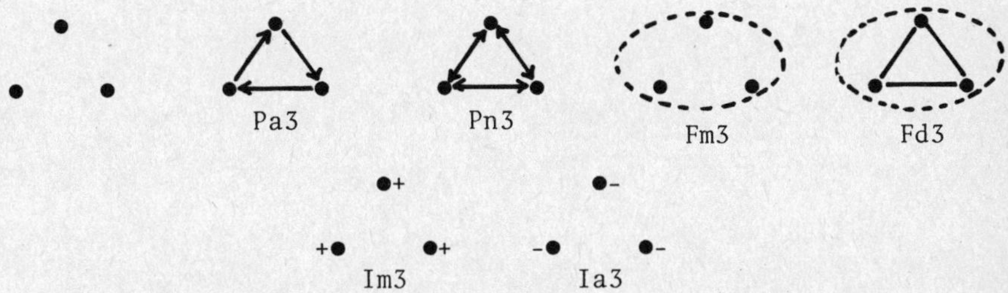

and subspace W invariant under A_3. The group S_3 is generated by $\alpha = (12)$ over A_3. The same argument as in the tetragonal case shows that $H^1(S_3, k^3/W) = 0$ for W = (000)(111), but is determined by colouring of a single vertex for W = (000) or (000)(011)(101)(110) by the above propositions together with the accompanying remarks. There are therefore 10 further

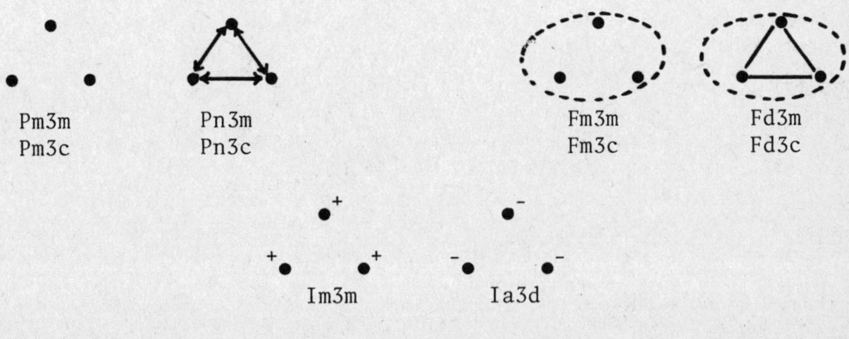

isomorphism classes.

6.7. CUBIC GROUPS FOR n=3

The remaining 3-dimensional space groups may be determined directly using the information provided in the last example. We have to calculate the cohomology groups $H^1(H,R^n/T)$ for $H = 432$, $\bar{4}3m$, 23 and $T = P,I,F$. In each case H can be generated by a subgroup H_1 of index 3, with generator θ rotation through $2\pi/3$, and by an additional generator α. We choose coordinates so that θ maps (a,b,c) to (c,a,b). A shift of origin in the plane perpendicular to (1,1,1) can be used to show that each element of $H^1(H,R^3/T)$ can be represented by a cocycle w with $w(\theta) = 0$; an argument used already in 6.3 to show that there is only one space group in the arithmetic crystal class (R,3). Then the cocycle is completely determined by the values of $w(\alpha)$; these are restricted to values consistent with the relations between α and θ, and are determined modulo changes of origin by a vector (x,x,x).

H = 432. Let $w(\alpha) = (a,b,c)$ where α maps (a,b,c) to (a,-c,b). Then

$$\alpha^4 = \iota \Longrightarrow (4a,0,0) \in T \Longrightarrow a = 0,\tfrac{1}{2},1,\tfrac{3}{2}$$

$$(\alpha\theta)^2 = \iota \Longrightarrow (a+c,0,a+c) \in T$$

change of origin $\Longrightarrow$ (a,b,c) changed to (a,b-2x,c)

The distinct cases are

(T = P)	$(0,0,0),\ (\tfrac{1}{2},0,\tfrac{3}{2}) \sim (\tfrac{3}{2},0,\tfrac{1}{2}),\ (1,0,1)$
(T = I)	$(0,0,0),\ (0,0,\tfrac{1}{2})$
(T = F)	$(0,0,0),\ (0,0,\tfrac{1}{2})$

where $\sim$ denotes equivalence under conjugation $\bar{\iota}$ and where b has been chosen 0 or $\frac{1}{2}$ as appropriate.

H = $\bar{4}3m$. Let $w(\alpha) = (a,b,c)$ where α maps (a,b,c) to (-a,c,-b). Then

$$\alpha^4 = \iota \Longrightarrow (0,0,0) \in T$$

$$(\alpha\theta)^2 = \iota \Longrightarrow (a-c,2b,c-a) \in T$$

change of origin $\Longrightarrow$ (a,b,c) changed to (a-2x,b,c-2x)

The distinct cases are

(T = P) $(0,0,0)$ $(0,1,0)$

(T = I) $(0,0,0)$ $(0,\frac{1}{2},1)$

(T = F) $(0,0,0)$ $(0,1,0)$

H = 23. Let $w(\alpha) = (a,b,c)$ where α maps (a,b,c) to $(a,-b,-c)$. Then

$$\alpha^2 = \iota \implies (2a,0,0) \in T \implies a = 0,1$$

$$(\alpha\theta)^3 = \iota \implies (d,-d,d) \in T \text{ where } d = a+c-b$$

change of origin $\implies$ (a,b,c) changed to $(a,b-2x,c-2x)$.

The distinct cases are

(T = P) $(0,0,0)$ $(1,0,1)$

(T = I) $(0,0,0)$ $(1,0,1)$

(T = F) $(0,0,0)$

As an interesting check on the results (and the basis of an alternative method of proof) we can use 6.1 to recover the already known space groups in geometric crystal class $m3m = 432 \times \bar{1} = \bar{4}3m \times \bar{1}$ and $m3 = 23 \times \bar{1}$. We may now complete the listings of 6.3 by the table showing the number of isomophism classes of space groups in each cubic arithmetic crystal class (T,H):

T = H =	P	I	F	Total
m3m	4	2	4	10
m3	3	2	2	7
432	3(+1)	2	2	7(+1)
$\bar{4}3m$	2	2	2	6
23	2	2	1	5

The final total is therefore 219 (+11) equivalence classes of space groups for n = 3.

6.8. HYPERCUBIC ORTHOGONAL GROUPS

The method of graphical calculation of cohomology outlined in 6.6 can be used to count isomorphism classes of orthogonal space groups for any subgroup $A \subseteq S_n$. The examples of 6.4 and 6.5 show that the numbers involved are very large when A is small. In this section we show that the numbers are very small when A is large, confirming the behaviour noticed already in 6.7 for n=3. We assume $n \geq 4$ and consider the alternating group A_n of order $n!/2$, and the symmetric group S_n of order n! The only subspaces $W \subset k^n$ invariant under $A = A_n$ or S_n are those corresponding to lattices P, U $(n \geq 4)$ and Z $(n > 4)$.

$\underline{A = A_n}$. Let $\beta = (123)$ be one generator for which, by 6.6, we assume $w(\beta) = 0$. The other generator can be chosen $\alpha = (34\ldots n)$ if n is odd, and $\alpha = (134\ldots n)$ if n is even. In both cases α has odd order and we consider the possible distinguished vertices on α-circuits. These may be altered by $x-\alpha x$ for $x \in k^n$ such that $x-\beta x = 0$, that is, for any x made up from vertices $4,\ldots,n$ (singly) or 1,2,3 (together).

<u>n odd</u>. Use x to ensure that $w(\alpha)$ has all or none of $3,4,\ldots,n$ distinguished, so $w(\alpha) = (abc\ldots c) = w(\alpha^{n-2})$ and the relation $\alpha^{n-2} = \iota$ implies $w(\alpha) \in W$.

<u>n even</u>. If vertex 3 together with an even number of $1,4,\ldots,n$ are distinguished then use x to ensure $w(\alpha) = (1b11\ldots1) = w(\alpha^{n-1})$ so that the relation $\alpha^{n-1} = \iota$ implies $w(\alpha) \in W$. In other cases use x to ensure $(abc0\ldots0)$ where either $c = 0$ or $c = a = 1$. Then $\alpha^{n-1} = \iota$ implies $(dbdd\ldots d) \in W$ where $d = a+c$, and $(\beta^{-1}\alpha)^{n-2} = \iota$ implies $(eeaa\ldots a) \in W$ where $e = b+c$ for n=4 and e=0 for $n > 4$. The three types of lattice give

$(T = P,\ n \geq 4)$: $b = d = 0,\ e = a = 0$ hence $w(\alpha) = 0$,

$(T = U,\ n \geq 4)$: $b = d = 0$ hence $w(\alpha) \in W$,

$(T = Z,\ n > 4)$: $b = d,\ e = a$ hence $w(\alpha) \in W$.

In all cases we conclude that $H^1(A_n, k^n/W) = 0$.

$\underline{A = S_n}$. The above calculation shows that $w \in H^1(S_n, k^n/W)$ may be represented by a cocycle zero on A_n. It follows that all transpositions $\gamma \in S_n$ have the same $w(\gamma) = x \in k^n/W$, and that x is invariant under the action of A_n. The three types of lattice give

$(T = P, n \geq 4)$: $x = (cc....c)$, $c = 0$ or 1

$(T = U, n \geq 4)$: $x = (co....o)$, $c = 0$ or 1

$(T = Z, n > 4)$: $x = (00....0)$

Therefore $H^1(S_n, k^n/W) = k$ in the first two cases and $H^1(S_n, k^n/W) = 0$ in the last case.

For $n > 3$ there are only two graphs invariant under the action of A_n on vertices. Hence the number of equivalence classes of graphs is (T = P)2, (T = U, n odd)2, (T = U, n even)1, (T = Z) 1. Combining these results with those on the cohomology of A we have the following numbers of isomorphism classes of space groups

Lattice	P, n=4	U or Z, n=4	P,n > 4	U,n > 4 (n odd)	U,n > 4 (n even)	Z,n > 4
$A = A_n$	2	1	2	2	1	1
$A = S_n$	4	2	4	4	2	1

7 Deformations

It is clear from 1.5 and 3.5 that a complete geometrical understanding of space groups must include not only discrete lists of isomorphism classes but also information about behaviour under continuous variation of structure [47].

7.1. THE SPACE OF n-DIMENSIONAL LATTICES

Each lattice $T \subset R^n$ of rank n is determined by an integral basis $\{t_1,...t_n\}$, that is by a non-singular matrix with real coefficients. The set of such matrices is denoted GL(n,R); it is a non-compact manifold of dimension n^2. Two bases determine the same lattice if they differ by multiplication by a non-singular matrix with integer coefficients, i.e. by an element of GL(n,Z). Thus the set of all T may be identified with the quotient set GL(n,R)/GL(n,Z), which is again a non-compact manifold of dimension n^2.

Proposition The space GL(n,R)/GL(n,Z) is connected.

Proof The space GL(n,R) has two components defined by the continuous map det : GL(n,R) $\rightarrow$ $\{\pm 1\}$. By changing t_1 to $-t_1$ if necessary we can always ensure that T is represented by $\{t_1,...,t_n\}$ in the component

$$SL(n,R) = \{\phi \in GL(n,R);\ \det \phi = +1\}.$$

Since SL(n,R) is connected so is the space of lattices; in fact the latter may be regarded as the quotient space SL(n,R)/SL(n,Z).

For many purposes it is convenient to identify lattices of the same shape whatever their relative magnitude or position in R^n.

Definition The lattices T,T' are *of the same shape* if there exists an orthogonal transformation β and a positive real number c such that $cT' = \beta T$. The set of equivalence classes is denoted L_n.

More formally we may write L_n as a *double quotient space* $R_+ \times O(n)\backslash GL(n,R)/GL(n,Z)$. The quotient $GL(n,R)/GL(n,Z)$ gives L_n a natural topology. Thus L_n is a connected non-compact space for $n > 1$.

Examples For n=1 the space L_1 consists of a single point. For n=2 the space L_2 consists of the non-compact region illustrated in 1.5. Note that there is a natural compactification of L_2 obtained by adjoining L_1. We

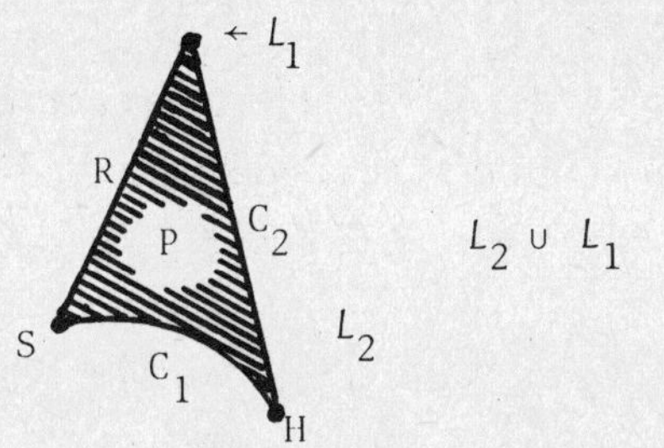

S = square
H = hexagon
R = rectangle
C = centred rectangle or diamond

shall prove in 7.2 that this is typical of low values of n.

If T,T' have the same shape then their symmetry groups G(T),G(T') are conjugate as subgroups of O(n). By a slight abuse of language we call L_n the *space of* n-*dimensional lattices*, write T for a point of L_n, and write G(T) for (the conjugacy class of) the symmetry group of T.

If $T \in L_n$ and $\alpha \in GL(n,R)$ then it is always true that $G(\alpha T)$ contains $\alpha G(T)\alpha^{-1} \cap O(n)$. Suppose $\delta \in G(\alpha T)$ so that $\alpha^{-1}\delta\alpha$ maps T to T. If α is sufficiently close to the identity then $G(\alpha T) \subseteq G(T)$ and $\alpha^{-1}\delta\alpha$ is close to an orthogonal map from T to T. Therefore $\alpha^{-1}\delta\alpha$ is orthogonal, and $G(\alpha T) \subseteq \alpha G(T)\alpha^{-1} \cap O(n)$. This proves a result used already in the proof of the theorem of 3.5: if α is sufficiently close to the identity then

$$G(\alpha T) = \alpha G(T)\alpha^{-1} \cap O(n).$$

In particular if an orthogonal transformation γ lies in both G(T) and $G(\alpha T)$ then $\gamma = \alpha\gamma\alpha^{-1}$, for both γ and $\alpha\gamma\alpha^{-1}$ map T to T and α is close to the identity.

Theorem Let G be a finite subgroup of O(n) and $L_n(G)$ the subspace of L_n consisting of lattices T with G(T) = G. Then $L_n(G)$ is an analytic manifold.

Proof Suppose $T \in L_n(G)$. For α sufficiently close to the identity $G(\alpha T) = G(T)$ if and only if $\gamma\alpha = \alpha\gamma$ for all $\gamma \in G(T)$, that is α lies in the centraliser

$$C(G,GL(n,R)) = \{\alpha \in GL(n,R);\ \gamma\alpha = \alpha\gamma \text{ for all } \gamma \in G\}.$$

Therefore a neighbourhood of T in $L_n(G)$ is homeomorphic to a neighbourhood of the identity element in the quotient space C(G,GL(n,R))/C(G,O(n)), or equivalently C(G,SL(n,R))/C(G,SO(n)). These spaces are analytic manifolds, because the equations $\gamma\alpha = \alpha\gamma$ are linear in the n^2 coefficients of α.

Remarks (i) We have actually shown more: the closure of $L_n(G)$ in L_n consists of $L_n(G)$ together with all spaces $L_n(G')$ with $G \subset G'$.

(ii) The example G = mm for n=2 shows that $L_n(G)$ need not be connected. Examples for n=3 show that moreover the closure of $L_n(G)$ need not be connected (see [47]).

7.2. THE SPACE OF REDUCED LATTICES

By 5.1 every lattice $T \subset R^n$ admits a reduced basis $\{t_1,\ldots,t_n\}$ with associated numbers $s_{ij} = (t_i,t_j)$. The latter determine a n **x** n symmetric matrix, or alternatively a vector $s = (s_{ij}) \in R^{\frac{1}{2}n(n+1)}$. If lattices T,T' have the same shape then the corresponding vectors s,s' differ only by a positive scalar multiple. We therefore assume

$$\sum_{i=1}^{n} s_{ii} = 1.$$

The vectors s which occur all lie within a closed bounded set S_n in the $(\frac{1}{2}n(n+1) - 1)$-dimensional hyperplane defined by the above equation; this follows already from the inequalities $\pm 2s_{jk} \leq s_{ij} \leq s_{kk}$ and $0 \leq s_{jj}$ of 5.1. The full description of the set S_n is by the inequalities

$$0 \le s_{ii} \le s_{jj} \le 1 \quad \text{for } 1 \le i \le j \le n$$

$$(*) \qquad 2\sum_i r_i s_{ik} \le \sum_i r_i^2 s_{ii} + 2\sum_{i<j} r_i r_j s_{ij} \quad \text{for } 1 \le i \le j \le k-1$$

and all integers r_i, r_j

Not all points of S_n arise from reduced bases of lattices of rank n. If $s \in S_n$ does arise from a reduced basis $\{t_1,\ldots,t_n\}$ then the fact that $\{t_1,\ldots,t_k\}$ is primitive implies

$$\Delta_k = \begin{vmatrix} s_{11} & s_{12} & \cdots & s_{1k} \\ s_{12} & s_{22} & & \\ \vdots & & & \\ s_{1k} & & & s_{kk} \end{vmatrix} \neq 0$$

Conversely if each such determinant for $k = 1,\ldots,n$ is non-zero we can construct a reduced basis. For suppose $t_1,\ldots,t_k$ have been found with $(t_i,t_j) = s_{ij}$ for $1 \le i \le j \le k$. Then we may seek t_{k+1} in the form $e_{k+1}+v$ where v lies in the space spanned by $t_1,\ldots,t_k$ and e_{k+1} is perpendicular to this space. Then

$s_{j,k+1} = (t_j,t_{k+1}) = (t_j,v) = \sum_{i=1}^{k} x_i s_{ij}$ if $v = \sum_{i=1}^{k} x_i t_i$. The non-vanishing of the determinant Δ_k shows that the numbers $s_{j,k+1}$ determine the coefficient coefficients x_i uniquely; the value of $s_{k+1,k+1}$ then determines $|e_{k+1}|^2 = s_{k+1,k+1} - |v|^2$ which is positive by the non-vanishing of the determinant Δ_{k+1}.

Definition The open dense subset S_n^o of S_n, defined by the non-vanishing of the above determinants, is called the *space of reduced n-dimensional lattices*.

There is a natural map $S_n^o \to L_n$ which is continuous and surjective. It comes from identifications in S_n^o corresponding to the various choices that may be made during the construction of the reduced basis of a lattice. These are

(i) if $\varepsilon_i = \pm 1$ are choices of sign for $i = 1,\ldots,n$ then identify (s_{ij}) with $(\varepsilon_i\varepsilon_j s_{ij})$;

(ii) if for some fixed k and choice of integers r_i, r_j there is an equality

$$2\sum_i r_i s_{ik} = \sum_i r_i^2 s_{ii} + 2\sum_{i<j} r_i r_j s_{ij}$$

where summations run from 1 to k-1 then alter all $s_{k\ell}$ by $\sum_i r_i s_{i\ell}$ for $\ell = 1,\ldots,k-1,k+1,\ldots,n$;

(iii) if for some fixed $j \le k$ there is an equality $s_{jj} = s_{kk}$ then interchange s_{ij} and s_{ik} for $i = 1,\ldots,n$.

These three kinds of identification correspond to the choices (i) t_i or $\varepsilon_i t_i$? (ii) t_k or $t_k - \sum_i r_i t_i$? (iii) t_j or t_k? Since all three are defined in terms of the numbers s_{ij} they make sense on S_n, not just on S_n^o. We denote the quotient under the identifications by Q_n and have

<u>Theorem</u> There is a compact topological space Q_n which contains L_n as an open dense subset.

We see that Q_n is not a manifold. There is however a surjective continuous map $S_n \to Q_n$ where S_n is a manifold with boundary (in fact a closed disc of dimension $\frac{1}{2}n(n+1)-1 = \frac{1}{2}(n+2)(n-1)$).

Points of Q_n which are not in L_n correspond to *singular* lattices. For example the equations $s_{1j} = 0$ for $j = 1,\ldots,n$ define a subspace of Q_n which is homeomorphic to Q_{n-1}. It corresponds to *singular* reduced bases $\{0,t_2,\ldots,t_n\}$. Thus Q_n contains a disjoint union $L_1 \cup L_2 \cup \ldots\ldots \cup L_n$.

<u>Theorem</u> If $n \le 4$ then $Q_n = L_1 \cup L_2 \cup \ldots\ldots \cup L_n$.

<u>Proof</u> Consider the determinants

$$\Delta_k = \begin{vmatrix} s_{11} & s_{12} \cdots s_{1k} \\ s_{12} & s_{22} \\ \vdots & \\ s_{1k} & & s_{kk} \end{vmatrix}$$

If $\Delta_1,\ldots,\Delta_k$ are non-zero then we may seek $t_{k+1} = e_{k+1}+v$ as above. The vectors e_i form an orthogonal basis for the vector space spanned by $t_1,\ldots,t_k$ and we may assume $v = \sum_{i=1}^{k} y_i e_i$ with $0 \le y_i \le \frac{1}{2}$. Then

$$|v|^2 \le \sum_{i=1}^{k} \tfrac{1}{4}|e_i|^2 \le \sum_{i=1}^{k} \tfrac{1}{4}|t_i|^2 = \tfrac{1}{4}\sum_{i=1}^{k} s_{ii} \quad .$$

is strictly less than $s_{k+1,k+1}$ for $k < 4$. It follows that $|e_{k+1}| > 0$ and hence Δ_{k+1} is non-zero. Therefore the only way in which there can be a point in $\mathcal{Q}_n$ which is not in L_n is if $\Delta_1 = 0$, that is the point lies in the subspace homeomorphic to $\mathcal{Q}_{n-1}$.

Remarks (i) Compare the examples in 5.1. (ii) The theorem is not true for $n > 4$ and it is then much less clear what is the *natural* compactification of L_n. This fact represents the major difference between crystallography in low dimensions and crystallography in arbitrary dimensions (compare [24], [47], [56]).

7.3. EXAMPLES FOR n=3

The space L_3 can be analysed very explicitly (see [47]). Certain subspaces of L_3 are of particular interest. In this section we describe the closure of $L_3(G)$ for G = mmm. For convenience we choose a preferred axis and let c be the length of the shortest non-zero vector (if any) of T along this axis. Then T determines (i) a point in L_2(mm) or in the corresponding space of orthogonal black and white axes, and (ii) a value of c. The points (i) lie along five one-dimensional manifolds. In the diagrams unmarked vertices correspond to *singular* lattices, that is to copies of L_1 and S,H,R,C have the same meaning as in 7.1, with colouring ignored in the black and white case. Now consider variation of the parameter c from c=0 (corresponding to

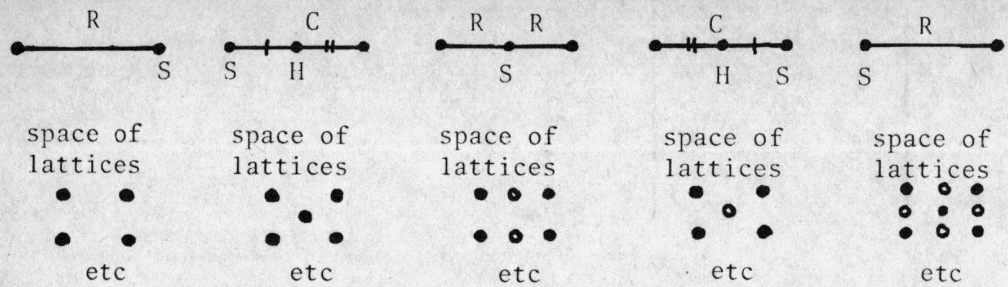

a copy of L_1). For particular values of c the symmetry of the lattice T may be larger than mmm. Such points are marked

P,I,F cubic symmetry, points of L_3(m3m)

---------- hexagonal symmetry, points of L_3(6/mmm)

 tetragonal symmetry, points of L_3(4/mmm)

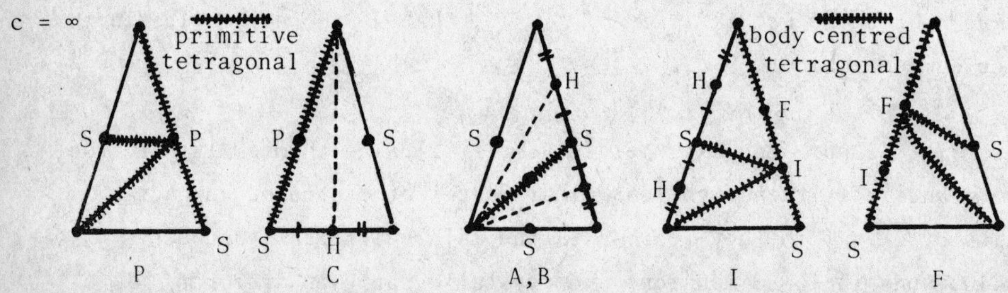

Now consider the effect of permutation of the three coordinates. It is to identify the three components of the faces marked P,I,F and to identify the two components of the face marked A,B with the face marked C. The result is the pair of discs

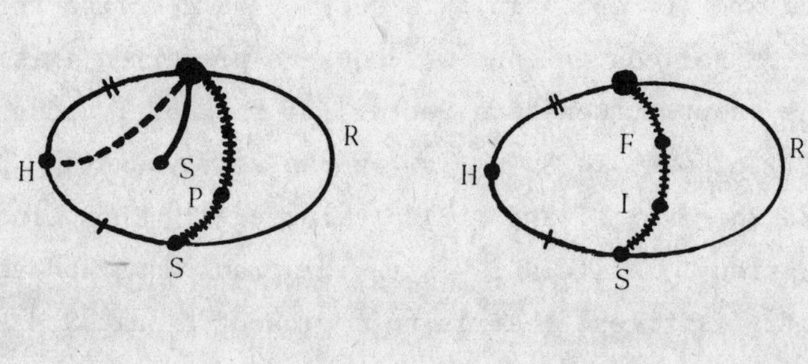

in which boundary points do not lie in L_3. Regarded as a subspace of of $L_1 \cup L_2 \cup L_3$, the remaining identifications of 2-dimension lattices give a cross-cap. This is the closure of L_3(mmm) in $L_1 \cup L_2 \cup L_3$.

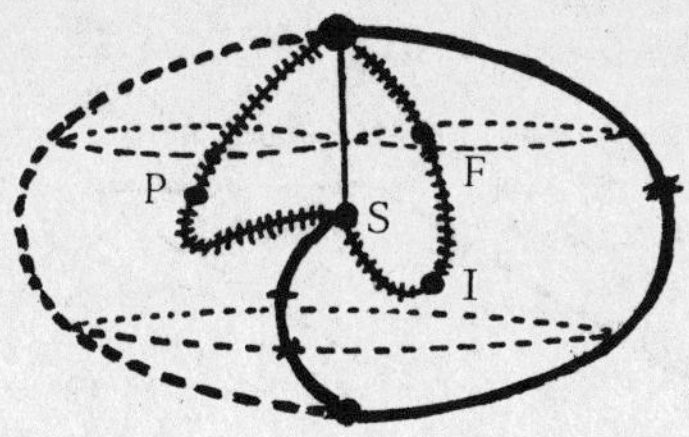

Notice that the closure of L_3(4/m mm) in L_3 is not connected, but the closure in the above space, which is part of $L_1 \cup L_2 \cup L_3$, is connected. On the other hand L_3(m3m) consists of isolated points and its closure is never connected.

To obtain the whole space $L_1 \cup L_2 \cup L_3$ it is necessary to attach further cells to the cross-cap. Attachment of the 2-cell giving the remaining points of L_2 to the 1-cell marked on the diagram with a solid line gives a 2-complex X with the homotopy type of the wedge of two spheres. Attachment of the 3-cells corresponding to primitive and body centred monoclinic lattices *fills in* the two spheres of X, giving a contractible 3-complex which represents all monoclinic lattices in R^3. Finally attachment of the 1-cell corresponding to rhombohedral lattices (which passes through P,I,F,H on the diagram) and of the 5-cell corresponding to triclinic lattices gives the complete space $L_1 \cup L_2 \cup L_3$. For details see [47].

Similar results may be obtained for n=4, using the information in [8] or [10].

7.4. TWINNING

The deformation of the lattice T, described above, is of course not the only way in which the structure of a space group can vary. To obtain useful models for describing variation of structure when n=3 it is necessary to pay regard to the position of the actual atoms, not merely of the translation vectors in the lattice, and to incorporate singularities, that is points in the neighbourhood of which symmetry is broken.

An important special case is that of two lattices T_1,T_2 with common intersection $T_1 \cap V$, $T_2 \cap V$ with a hyperplane V. A satisfactory model should describe how T_1 *deforms* into T_2 as, under some reasonably hypotheses, the hyperplane moves across R^n. Any lattice T of rank n has a well defined

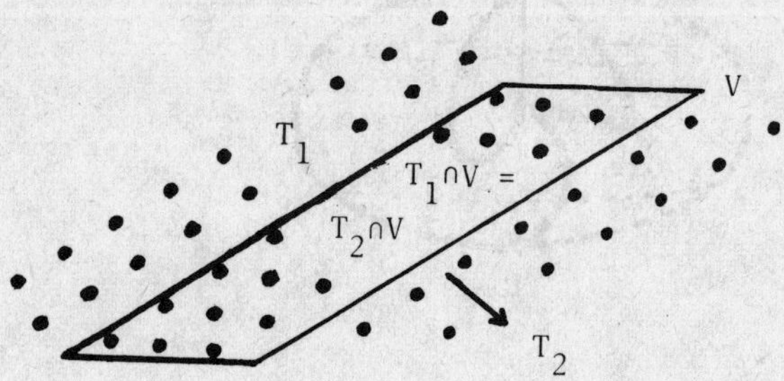

volume, defined as the determinant of the n × n matrix with columns $t_1,\ldots,t_n$ where $\{t_1,\ldots\ldots,t_n\}$ is some choice of integral basis. Another choice of integral basis will yield the same volume up to sign. We write

$$\text{Vol } T = |\det(t_1,\ldots,t_n)|$$

<u>Definition</u> A *twinning* of the arithmetic classes (T_1,H) and (T_2,H) consists of

(i) a hyperplane $V \subset R^n$

(ii) a choice of minimal lattices T_1,T_2 such that Vol T_1 = Vol T_2 and $T_1 \cap V = T_2 \cap V = T$ is a lattice of rank n-1.

Thus twinned lattices admit an action by the same (not merely by a conjugate) point group H, and share a common sublattice of codimension 1.

<u>Theorem</u> Let (T_1,H) and (T_2,H) be arithmetic crystal classes in the same geometric crystal class. Suppose there exists a hyperplane $V \subset R^n$ such that (i) $T_1 \cap V = T_2 \cap V$ is a lattice T of rank n-1, and (ii) V is invariant under the action of H. Then there is a twinning of (T_1,H) and (T_2,H).

<u>Proof</u> The invariance of V implies that an arithmetic crystal class (T,H) contains actions $(\phi T,H)$ where ϕ is the linear map defined by multiplication by a scalar $c > 0$ along the axis perpendicular to V, and the identity on V.

Replace T_1, T_2 by $\phi_1 T_1$, $\phi_2 T_2$ with scalars chosen to ensure Vol T_1 = Vol T_2.

<u>Examples</u> (i) Let T_1, T_2 be tetragonal lattices P,I for n=3 and V the plane perpendicular to the axis of the rotation of order 4 in $H = 4, \bar{4}, 4/m, 4mm, 422, \bar{4}2m$ or $4/m\ mm$.

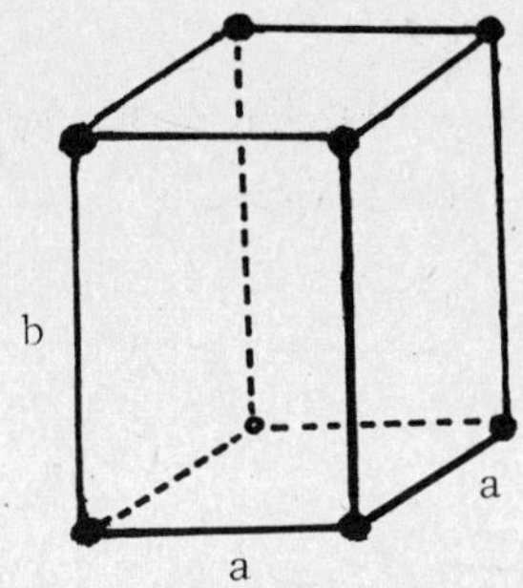

Lattice P
Volume a^2b

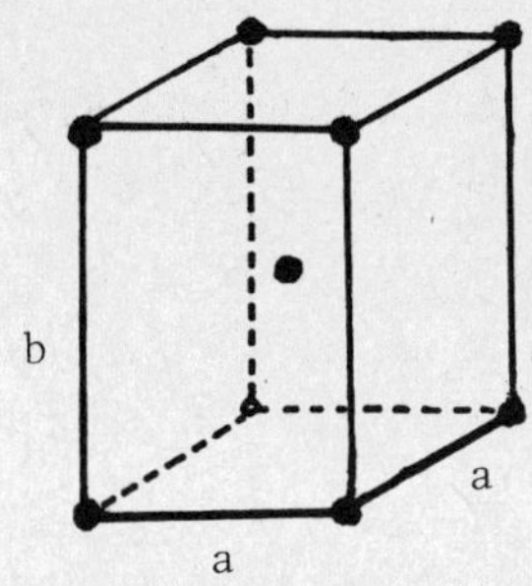

Lattice I
Volume a^2b

There is a twinning between (P,H) and (I,H) in which P has lattice parameters a,b, and I has lattice parameters a,2b.

(ii) Let T_1, T_2 be hexagonal lattices P,R for n=3 and V the plane perpendicular to the axis of rotation of order 3 in $H = 3, \bar{3}, 3m, 32$ or $\bar{3}m$. There is a twinning between (T_1, H) and (T_2, H). Note that in this case $G(T_1) \neq G(T_2)$; thus twinning should be regarded as a relation between arithmetic crystal classes in the same *family*, rather than as a relation between lattices in the same system.

(iii) Consider the cubic lattices P,I,F for n=3. Now there is no plane invariant under the action of any of the groups $H = 23, m3, 432, \bar{4}3m$ or $m3m$. These lattices can be regarded as specialisations of those in (i): thus tetragonal P with a = b gives cubic P with volume b^3, tetragonal I with a = b gives cubic I with volume $\frac{1}{2}b^3$, and tetragonal I with $a = b/\sqrt{2}$ gives cubic F with volume $\frac{1}{4}b^3$. No twinning can exist between (T_1, H) and (T_2, H) with T_1, T_2 a distinct pair from P,I,F: for if it did equality of volumes would imply that T_1, T_2 have lattice parameters b_1, b_2 with $b_1 : b_2$ equal to $\sqrt[3]{\frac{1}{4}}$, $\sqrt[3]{\frac{1}{2}}$, $\sqrt[3]{2}$ or $\sqrt[3]{4}$, and comparison of the lengths of vectors in T_1, T_2 would show that there is no common sublattice $T_1 \cap V = T_2 \cap V$ of rank 2.

These examples show that the possibilities for twinning between arithmetic crystal classes of a given family are intimately connected with the partition of n determined by the family as in 4.4: the more components in the partition, the more possibilities for twinning. The same phenomenon occurs when the hyperplane V is not invariant under H, as in the following examples.

<u>Examples</u> (i) Let T_1, T_2 be the orthogonal lattices P,C for n = 2, and let H = mm. This group is generated by reflections in two perpendicular lines each containing shortest non-zero vectors. Let a_1, b_1 be the lengths of the two vectors for T_1 and a_2, b_2 the lengths for T_2. Then

$$\text{Vol } T_1 = a_1 b_1 \quad ,$$
$$\text{Vol } T_2 = \tfrac{1}{2} a_2 b_2 \ .$$

The lattice $T = T_1 \cap V = T_2 \cap V$ given by a twinning of (T_1, H) and (T_2, H) must contain a vector of the form $(r_1, s_1) \in T_1$ with $\text{hcf}(r_1, s_1) = 1$. Here r_1, s_1 are integers and coordinates are with respect to the shortest non-zero vectors of T_1 on reflection lines. Similarly T must contain $(r_2, s_2) \in T_2$ with $\text{hcf}(r_2, s_2) = 1$. Then (r_1, s_1) is a shortest non-zero vector in $T_1 \cap V$, and either (r_2, s_2) or, if both r_2 and s_2 are odd, $(\tfrac{1}{2}r_2, \tfrac{1}{2}s_2)$ is a shortest non-zero vector in $T_2 \cap V$.

<u>Lemma</u> With the above notation $a_1 r_1 = a_2 r_2$ and $b_1 s_1 = b_2 s_2$.

<u>Proof</u> It is sufficient to show that (r_2, s_2) is a shortest non-zero vector in $T_2 \cap V$. If not then $(\tfrac{1}{2}r_s, \tfrac{1}{2}s_2) \in T_2 \cap V$, that is r_2 and s_2 are both odd and $r_1 a_1 = \tfrac{1}{2} r_2 a_2$, $b_1 s_1 = \tfrac{1}{2} b_2 s_2$. Now equality of volumes gives $2a_1 b_1 = a_2 b_2$ and hence $r_1 s_1 = \tfrac{1}{2} r_2 s_2$ is not an integer. This is the required contradiction.

<u>Proposition</u> With the above notation twinning can occur with $(r_1, s_1) = (4p, 2q+1)$ and $(r_2, s_2) = (2p, 2q+1)$ for any integers p,q such that p and 2q+1 are coprime.

<u>Proof</u> By the lemma, and equality of volumes,

$$r_1 s_1 = 2r_2 s_2$$

The proof of the lemma shows that r_2 and s_2 have opposite parity, so assume $r_2 = 2p$ and $s_2 = 2q+1$. Then p and 2q+1 are coprime. Certainly $(r_1,s_1) = (4p,2q+1)$ satisfies all conditions (for particular values of p and q there may be other solutions for r_1,s_1).

The diagram illustrates the case p=1, q=0.

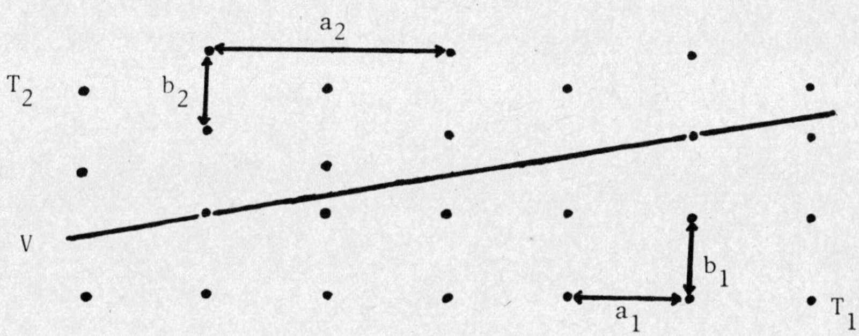

(ii) The group mm did not play an essential role in the above computation; therefore twinning between (P,m) and (C,m) occurs in the same way. The same argument for n=3 gives examples of twinning between the orthogonal classes

(P,mmm) and (C,mmm), (P,2mm) and (C,2mm), (P,222) and (C,222)

(F,mmm) and (I,mmm), (F,2mm) and (I,2mm), (F,222) and (I,222),

in which V is not invariant under the action of H.

(iii) Similarly, if • and ∘ represents points with third coordinate 0 and $\frac{1}{2}$ mod Z respectively, then the diagram

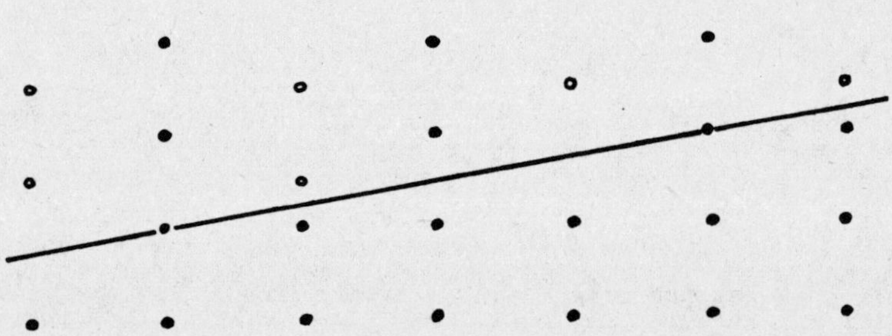

describes twinning of (P,mmm) and (I,mmm), (P,2mm) and (I,2mm), (P,222) and (I,222).

The twinning of arithmetic crystal classes can also be formulated in terms of colour symmetry, when V is a given hyperplane not invariant under H. For then $T = T_1 \cap V = T_2 \cap V$ defines a lattice

$$T_o = \{\theta t;\ \theta \in H,\ t \in T\}$$

which is of rank n and contained in both T_1 and T_2. The colour lattices $T_o \subseteq T_1$ and $T_o \subseteq T_2$ need not be identical (as the examples above show) but they must share many properties: T_i/T_o is cyclic and generated by an equivalence class $[t_i]$, where the perpendicular distance d_i from t_i to V is chosen as short as possible, and where $d_1 = d_2$ by equality of volumes. For instance the examples of coloured lattices given at the end of Chapter Five are well known to physicists interested in twinning (see further Chapter Ten of [30]).

Appendix

The study of mathematical crystallography was begun by physicists. Viewed as a part of pure mathematics it lies between algebra and geometry. The terminology fashionable in the latter subjects has changed rapidly in the last 20 years. Hence results and terminology which are familiar to one reader may be unfamiliar to another, and a brief guide to some topics of this kind may be helpful.

1. EXACT SEQUENCES OF GROUPS

Most groups will be groups of symmetries with multiplication defined by composition, so we write $\iota \in G$ for the identity element and 1 for the group $\{\iota\}$ consisting only of ι.

Definition A sequence $\ldots \xrightarrow{\alpha_i} G_i \xrightarrow{\alpha_{i+1}} G_{i+1} \longrightarrow \ldots$ of groups G_i and homomorphisms $\alpha_i : G_{i-1} \longrightarrow G_i$ is *exact* if $\mathrm{im}\,\alpha_i = \ker \alpha_{i+1}$ for each i.

Thus the sequence

$$1 \longrightarrow K \xrightarrow{\alpha} F \xrightarrow{\beta} H \longrightarrow 1$$

is exact if $\{\iota\} = \ker \alpha$, $\mathrm{im}\,\alpha = \ker \beta$, $\mathrm{im}\,\beta = H$. This means simply that α is injective (a monomorphism), and can be identified with the subgroup $\alpha(K)$ of F, that β is surjective (an epimorphism) with kernel $\alpha(K)$, and hence that H can be identified with the quotient group F/K.

The reason for using the exact sequence notation rather than quotient groups is that in geometric examples H is a concrete group rather than a collection of cosets. Examples with $K = \{\iota,\psi,\psi^2\} = C_3$ and $H = \{\iota,\theta\} = C_2$ are

$$1 \longrightarrow C_3 \xrightarrow{\alpha} C_6 \xrightarrow{\beta} C_2 \longrightarrow 1 \qquad\qquad 1 \longrightarrow C_3 \xrightarrow{\alpha} S_3 \xrightarrow{\beta} C_2 \longrightarrow 1$$

C_6 is cyclic group generator ϕ — S_3 is symmetric group of permutations

$$\alpha(\psi) = \phi^2 \qquad\qquad \alpha(\psi) = (123)$$

$$\beta(\phi) = \beta(\phi^3) = \beta(\phi^5) = \theta \qquad\qquad \beta((12)) = \beta((23)) = \beta((31)) = \theta$$

$$\beta(\iota) = \beta(\phi^2) = \beta(\phi^4) = \iota \qquad\qquad \beta(\iota) = \beta((123)) = \beta((321)) = \iota$$

Definition The exact sequence

$$1 \longrightarrow K \xrightarrow{\alpha} F \xrightarrow{\beta} H \longrightarrow 1$$

is *semi-split* if there exists a homomorphism $\sigma : H \to F$ such that $\beta\sigma = id_H$, and *split* if moreover there exists a homomorphism $\rho : F \to K$ such that $\rho\alpha = id_K$ and $\rho\sigma(\theta) = \iota$ for all $\theta \in H$.

If the exact sequence $1 \longrightarrow K \xrightarrow{\alpha} F \xrightarrow{\beta} H \longrightarrow 1$ is split then there is an isomorphism between F and $K \times H$ defined by maps

$$F \longleftrightarrow K \times H$$

$$\phi \longmapsto \rho(\phi),\ \beta(\phi)$$

$$\alpha(\psi)\sigma(\theta) \longleftarrow\!\shortmid\ \psi,\theta$$

and so every element of F may be written as a pair of elements from K and H. For example $C_6 = C_3 \times C_2$ with $\phi = (\psi,\theta)$, $\phi^2 = (\psi^2,\iota)$, $\phi^3 = (\iota,\theta)$, $\phi^4 = (\psi,\iota)$, $\phi^5 = (\psi^2,\theta)$, and $\phi^6 = \iota = (\iota,\iota)$, but $S_3 \neq C_3 \times C_2$. The first example above is split; the second is only semi-split.

If K is an abelian group, written additively with identity element $0 \in K$, then the above exact sequence is written

$$0 \longrightarrow K \xrightarrow{\alpha} F \xrightarrow{\beta} H \longrightarrow 1$$

and there is action of H on K defined by

$$\theta \cdot k = fkf^{-1} \text{ where } \beta(f) = \theta.$$

Note that $\theta \cdot k$ is well defined (if f' is another choice then $f' = fx$ for some $x \in k$ and so $f'kf'^{-1} = fxkx^{-1}f^{-1} = fkf^{-1}$) and $\theta \cdot (k+\ell) = \theta \cdot k + \theta \cdot \ell$ (because $fk\ell f^{-1} = fkf^{-1}f\ell f^{-1}$).

2. SYMMETRIC MATRICES

In the proof of a theorem in 2.3 the following result was quoted.

Theorem Let ω be a positive definite symmetric matrix. There is a maxtrix α such that $\alpha^2 = \omega$ and $\alpha = a_o + a_1\omega + \ldots + a_m\omega^m$ for some integer m and real numbers $a_o, a_1, \ldots, a_m$.

Proof The hypothesis implies $\omega = \omega'$ and $x'\omega x > 0$ for all non-zero $x \in R^n$. If $\lambda = \mu + i\nu$ is a complex eigenvalue for ω with eigenvector $z = x+iy$ then $\omega \cdot z = \lambda \cdot z$ implies

$$\lambda \bar{z}'z = \bar{z}'\omega z = \bar{z}'\omega' z = \overline{(\bar{z}'\omega z)'} = \overline{(\lambda \bar{z}'z)'} = \bar{\lambda}\bar{z}'z$$

(where - and ' denote complex conjugate and transpose respectively) so that $\nu = 0$ and $\mu > 0$. Therefore the characteristic polynomial for ω has the form

$$P(\lambda) \equiv (a-\lambda)^r (b-\lambda)^s \ldots .$$

where a,b,.... are positive real numbers. By the Cayley-Hamilton theorem the matrix ω satisfies its own characteristic equation. It is therefore sufficient to find a polynomial $Q(\lambda)$ of degree $m < n$ such that $P(\lambda)$ divides $Q^2(\lambda) - \lambda$ and to define $\alpha = Q(\omega)$ for then

$$\alpha^2 - \omega = Q^2(\omega) - \omega = P(\omega)\ R(\omega) = 0.$$

To find the polynomial $Q(\lambda)$ write $P_a(\lambda) = \dfrac{P(\lambda)}{(a-\lambda)^r}$, $P_b(\lambda) = \dfrac{P(\lambda)}{(b-\lambda)^s}$, ... and seek polynomials $A(\lambda)$, $B(\lambda)$,.... of degree $<$ r, s,.... such that $(a-\lambda)^r$, $(b-\lambda)^2$,... divide $A^2(\lambda)\ P_a^2(\lambda) - \lambda$, $B^2(\lambda)\ P_b^2(\lambda) - \lambda$, ... respectively. Then

$$Q(\lambda) = A(\lambda)\ P_a(\lambda) + B(\lambda)\ P_b(\lambda) + \ldots .$$

The polynomials $A(\lambda)$, $B(\lambda)$,.... are defined uniquely up to sign since, writing $y = a-\lambda$ and $P_a(a-y) = d_o+d_1y+....$, the condition

$$y^r \text{ divides } A^2(a-y)\, P_a^2(a-y) + y-a$$

implies that the coefficients of

$$A(a-y) = c_o+c_1y+....+c_{r-1}y^{r-1}$$

satisfy

$$d_o^2c_o^2 - a = 0 \quad (\text{determining } \pm c_o \text{ since } a > 0)$$

$$2d_o^2c_oc_1 + 2c_o^2d_od_1+1 = 0 \quad (\text{determining } c_1)$$

$$2d_o^2c_oc_2 +... = 0 \quad (\text{determining } c_2)$$

....

3. TOPOLOGY OF O(n)

In 3.1 the fact that the orthogonal group is a compact topological space was used. The topology of $O(n)$ arises from the fact that the groups

$$O(n) = \{\theta;\ \theta'\theta = \iota\} \subset M_n$$

$$GL(n,R) = \{\theta;\ \theta \text{ invertible}\} \subset M_n$$

can be regarded as subsets of the ring M_n of all $n \times n$ matrices, and hence as subsets of n^2-dimensional euclidean space.

Explicitly $O(n)$ is the set of points $\theta = (x_{ij})$ in n^2-dimensional space which satisfy the $\frac{1}{2}n(n-1)+n$ equations

$$\sum_{j=1}^{n} x_{ij}x_{kj} = 0 , \quad 1 \le i < k \le n .$$

$$\sum_{j=1}^{n} x_{ij}^2 = 1 , \quad 1 \le i \le n.$$

Thus $O(n)$ is closed. It is also bounded since

$$\| \theta \|^2 = \sum_{i,j} x_{ij}^2 = n$$

and therefore O(n) is compact. If P is an arbitrary compact space then $P \times O(n)$ is compact (compare 3.1).

The map det: $O(n) \longrightarrow \{\pm 1\}$ is clearly continuous and gives an exact sequence

$$1 \longrightarrow SO(n) \longrightarrow O(n) \xrightarrow{\det} \{\pm 1\} \longrightarrow 1$$

which is semi-split: the homomorphism $\sigma : \{\pm 1\} \longrightarrow O(n)$ is given by letting $\sigma(-1)$ be a diagonal matrix β with an odd number of diagonal elements -1 and the remainder +1. Then $O(n) = SO(n) \cup \beta\, SO(n)$ consists of two disjoint copies of SO(n). In particular SO(n) is compact.

To decide when the sequence is split note that any homomorphism $\rho : O(n) \longrightarrow SO(n)$ with $\rho(\beta) = \iota$ also has $\rho(\alpha\beta\alpha^{-1}) = \iota$. Hence, if the sequence is split,

$$\alpha\beta = \beta\alpha \quad \text{for all } \alpha \in O(n).$$

This is a contradiction unless n is odd and $\beta = \bar{\iota}$. Therefore the sequence is not split for n even. For n odd the sequence is split and there is an isomorphism

$$O(n) \longrightarrow SO(n) \times \{\pm 1\}$$

(for the case n=3 compare 4.3).

If $\theta \in SO(n)$ then the equation $\theta'\theta = \iota$ implies that all eigenvalues of θ have the form $e^{i\omega}$ where $0 \le \omega < 2\pi$. There is therefore a choice of coordinates with respect to which the matrix θ becomes a block matrix with components 1 or

$$\begin{matrix} \cos\omega & \sin\omega \\ -\sin\omega & \cos\omega \end{matrix}$$

In such a case let θ_t be the block matrix with components 1 or

$$\begin{matrix} \cos \omega t & \sin \omega t \\ -\sin \omega t & \cos \omega t \end{matrix} .$$

We see that $\theta_o = \iota$, $\theta_1 = \theta$ giving a path from ι to θ. In the same way any two points of SO(n) may be joined by a path and therefore SO(n) is connected.

For low values of n the space SO(n) may be identified with better known topological spaces. If P^n and S^n denote n-dimensional real projective space and sphere respectively then

$$SO(2) = P^1 = S^1$$

$$SO(3) = P^3$$

$$SO(4) = P^3 \times S^3$$

In general SO(n) is a manifold of dimension $\frac{1}{2}n(n-1)$. Since any $\theta \in SO(n)$ can be regarded as a block matrix $(v_1 \ldots v_n)$ where v_i is a unit vector, there is a natural continuous onto map

$$\begin{matrix} SO(n) & \longrightarrow & S^{n-1} \\ (v_1 \ldots . v_n) & \longmapsto & v_1 \end{matrix}$$

for which each counter-image is homeomorphic to SO(n-1).

4. COHOMOLOGY OF GROUPS

In 2.4, 3.2, 3.4, 3.6, 6.1-6.8 use was made of arguments and terminology involving cohomology. There is no need to understand the full theory in order to follow the very special cases needed for crystallography, but the following summary may make the terminology seem more natural. Throughout H denotes a group with a *given* action on an abelian group T satisfying $\theta \cdot (k+\ell) = \theta \cdot k + \theta \cdot \ell$.

We consider the r-th cochain group $C^r(H,T)$. For r=0 it is T. For $r > 1$ it consists of all functions

$$w : H \times \ldots . \text{ (r factors) } \ldots . \times H \longrightarrow T$$

satisfying $w(\theta_1,\ldots,\iota,\ldots,\theta_r) = 0$. It is an abelian group with addition induced by the addition in T.

There are homomorphisms

$$\delta : C^r(H,T) \longrightarrow C^{r+1}(H,T)$$

defined by the formula

$$\delta w(\theta_1,\ldots,\theta_{r+1}) = w(\theta_1,\ldots,\theta_r) + \sum_{i=2}^{r+1} (-1)^{r-i} w(\theta_1,\ldots,\theta_{i-1}\theta_i,\ldots,\theta_{r+1}) + (-1)^{r+1}\theta_1 w(\theta_2,\ldots,\theta_{r+1}).$$

For r=0 there are no terms in the summation and the formula becomes $\delta t(\theta) = t-\theta t$. For r=1 there is one term in the summation and the formula becomes

$$\delta w(\theta_1,\theta_2) = w(\theta_1) - w(\theta_1\theta_2) + \theta_1 w(\theta_2) .$$

For all r it can be checked that $\delta w \in C^{r+1}(H,T)$ and that the element $\delta(\delta u) \in C^{r+1}(H,T)$ is zero for all $u \in C^{r-1}(H,T)$.

<u>Definition</u> $w \in C^r(H,T)$ is a *cocycle* if $\delta w = 0$, and a *coboundary* if $w = \delta u$ for some $u \in C^{r-1}(H,T)$. The r-th *cohomology group* of the action of H on T is the quotient

$$H^r(H,T) = \{w \in C^r(H,T);\ w \text{ cocycle}\}/\{w \in C^r(H,T);\ w \text{ coboundary}\}.$$

For r=0 there are no coboundaries and the definition becomes $H^0(H,T) = \{t \in T;\ t - \theta t = 0 \text{ for all } \theta \in H\}$. For r=1 it gives the set of all functions $w : H \to T$ satisfying $w(\iota) = 0$ and $w(\theta_1\theta_2) = w(\theta_1) + \theta_1 w(\theta_2)$ modulo those of the form $w(\theta) = t - \theta t$ for fixed $t \in T$. Clearly $H^r(H,T)$ is always an abelian group.

<u>Theorem</u> Let H be a finite group of order N and $w \in H^r(H,T)$. Then $Nw = 0$.

<u>Proof</u> Represent w by a particular cocycle $w \in C^r(H,T)$. We shall prove the

existence of $u \in C^{r-1}(H,T)$ such that $Nw = \delta u$. Let

$$u(\theta_1,\ldots,\theta_{r-1}) = \sum_{\theta\in H} w(\theta_1,\ldots,\theta_{r-1},\theta).$$

The fact that w is a cocycle implies

$$w(\theta_1,\ldots,\theta_r) + \sum_{i=2}^{r+1} (-1)^{r-i} w(\theta_1,\ldots,\theta_{i-1}\theta_i,\ldots,\theta_r,\theta) +$$

$$(-1)^{r+1}\theta_1 w(\theta_2,\ldots,\theta_r,\theta) = 0$$

and summation of these equations over all $\theta \in H$ gives

$$N\, w(\theta_1,\ldots,\theta_r) + \sum_{i=2}^{r} (-1)^{r-i}\, u(\theta_1,\ldots,\theta_{i-1}\theta_i,\ldots,\theta_r) -$$

$$u(\theta_1,\ldots,\theta_r) + (-1)^{r+1}\theta_1 u(\theta_2,\ldots,\theta_r) = 0$$

$$N\, w(\theta_1,\ldots,\theta_r) - \delta u(\theta_1,\ldots,\theta_r) = 0.$$

So $Nw = \delta u$ as required.

Corollary If T is finitely generated and H is finite then $H^r(H,T)$ is finite.

Proof To specify $w \in H^r(H,T)$ it is sufficient to specify the values $w(\theta_i)$ of a cocycle which represents w. Therefore $H^r(H,T)$ is a finitely generated abelian group in which each element has order dividing N.

Corollary (Divisibility theorem). Suppose that every element of T is divisible by the order of H. Then $H^r(H,T) = 0$.

Proof Because then $Nw = \delta u$ implies $w = \delta(\frac{1}{N}u)$.

If H acts on an exact sequence

$$0 \longrightarrow T_1 \xrightarrow{\alpha} T \xrightarrow{\beta} T_2 \longrightarrow 0$$

of abelian groups then there is an exact sequence

$$0 \longrightarrow H^o(H,T_1 \xrightarrow{\alpha^*} H^o(H,T) \xrightarrow{\beta^*} H^o(H,T_2) \xrightarrow{\delta^*} H^1(H,T_1) \longrightarrow \quad \ldots.$$

$$\ldots \longrightarrow H^{r-1}(H,T_2) \xrightarrow{\delta^*} H^r(H,T_1) \xrightarrow{\alpha^*} H^r(H,T) \xrightarrow{\beta^*} H^r(H,T_2) \longrightarrow \quad \ldots.$$

of cohomology groups. Special cases of such a cohomology exact sequence occur in 3.4 and 3.6. The homomorphisms

$$\delta^* : H^{r-1}(H,T_2) \longrightarrow H^r(H,T_1)$$

are defined as follows. If $u \in H^{r-1}(H,T_1)$ is represented by a cocycle then $u = \beta v$ for some v in $C^{r-1}(H,T)$ such that $\beta(\delta v) = 0$. Therefore $\delta v \in T_1$ is a cocycle representing an element of $H^r(H,T_1)$. The homomorphisms α^*, β^* arise directly from composition of cochains with α, β.

Historical and bibliographical note

The earliest work on n-dimensional space groups was by physicists in the case n = 3. It was pointed out by Fedorov [17] that the finite subgroups $H \subset O(3)$ had been enumerated many times: by Hessel [25] as early as 1829 and in the crystallographic case by Bravais and Gadolin. It follows that the *crystallographic restriction* was known by physicists long before Barlow (who in England is sometimes credited with the result and who did give a neat proof of it which has survived in some textbooks). The 14 possible lattices $T \subset R^3$ were enumerated by Bravais [6] correcting earlier work of Frankenheim.

The first mathematician to attempt a study of the group A of invertible affine transformations and of the canonical exact sequence of 2.1 appears to have been Rodrigues [41]. (A discussion of his work has recently been given by Gray [22]).

Further contributions were made by Möbius [35] but the central paper is that of Jordan [29] which was partly motivated by the crystallographic results of Bravais and which influenced all later work. Jordan listed certain subgroups of A and the list was completed by Sohncke [53] ten years later to embrace all space groups consisting of orientation preserving transformations. By 1880 the essential foundations of n-dimensional crystallography had been laid. On this basis Fedorov - motivated by his work for the Russian Geological Survey - and Schönflies - prompted by Klein - were able to build the complete description of 3-dimensional space groups.

It is sometimes said that the enumeration of the 230 3-dimensional groups was carried out independently by Barlow, Fedorov and Schonflies, and even that this marks one of the most remarkable instances of independent discovery on record. In fact this myth was fostered by Fedorov who in 1892 wrote (with reference to himself and Schönflies only since at this stage Barlow had hardly begun work on the topic)

"....an extremely surprising circumstance has come to light, *viz* a coincidence in the work of two researchers such as has, perhaps, never been observed in the history of science" [17].

The original papers show immediately that the circumstances are less surprising than appears at first sight. Firstly both Fedorov and Schönflies had certainly read the book of Sohncke [53] with its list of orientation preserving groups, and both had noticed that one group had been listed twice [44], [14]. Secondly both had almost certainly read the subsequent papers of Sohncke [54], [55] which showed the need to consider structures more general than those summarised by the corrected list of 65 orientation preserving groups. Thirdly neither Fedorov nor Schönflies succeeded in achieving a correct list at the first attempt; each had to make corrections as a result of comparison with the list of the other. The original list of Schönflies [45] consisted of 227 groups (four being omitted and one duplicated so that 227+4-1 = 230). The extract from the minutes of the St. Petersburg Mineralogical Society in [16] suggests that the original list of Fedorov consisted of 229 groups (two being omitted and one duplicated so that 229+2-1 = 230). As a result of correspondence between Fedorov and Schönflies all errors were eliminated, the two lists [15] [46] being set in one-one correspondence by Fedorov in the article [17] which includes the above quotation.

Only in 1894, after reading the work of Schönflies and the brief article [17] of Fedorov, did Barlow publish his own attempt [1] to understand the lists of Fedorov and Schönflies by direct deduction from the better known work of Sohncke. The comments [18] of Fedorov on this attempt were rather uncharitable:

".... unfortunate that Barlow didn't follow through how his results coincide with mine, which would have been very easy .... even though he had more than one complete list in front of him he missed out some groups ...."

The list of Barlow [1] consisted of 229 groups (four being omitted and three duplicated so that 229+4-3 = 230) and was followed by a pained correction [2]. So much for the myth of independent discovery. The real historical interest of the period following Jordan and Sohncke lies not in the apportionment of credit between Barlow, Fedorov and Schönflies but in the striking parallel development of the mathematical idea of considering orientation reversing transformations as well as orientation preserving transformations, and of the physical idea of considering arrays of atoms or molecules as well as arrays of solid building blocks (see Sohncke [55]).

The twenty years following the work of Fedorov and Schönflies saw the problem placed in a more general context. The work of Fricke and Klein on automorphic forms [21] dealt with the corresponding problem in hyperbolic space. The eighteenth of the 23 problems stated by Hilbert at the 1900 International Congress of Mathematicians [26] was primarily concerned with the extension to n-dimensional space of the work of Fedorov [19] on building up space from congruent polyhedra. It also stated explicitly the question: "is there only a finite number of essentially different kinds of groups of motions with a fundamental region in n-dimensional euclidean space". The question was answered almost immediately by Bieberbach [3] using earlier results of Jordan. The remaining results mentioned in Chapter Three are similarly mainly due to Bieberbach [4].

The enumeration of plane symmetry groups is of course implicit in the 3-dimensional work. It was undertaken explicitly by Fedorov [16] but published only in Russian. Apart from oblique references in Fricke and Klein [21] the next proofs (without details) are those of Niggli [37] and Polya [39]. The proof in Chapter One is based on Schwarzenberger [49] and has been adapted to the frieze pattern group case by Macdonald and Street [32]. For the black and white plane groups see Shubnikov [51] and Loeb [31].

Improved algorithms for the determination of 3-dimensional space groups were developed by Burckhardt in a series of papers [11], and in his influential book [12]. The related algorithm of Zassenhaus [61] allowed the possibility of computational determination of space groups with given arithmetic crystal class $H \subset GL(n,Z)$, a development which led to the complete determination of the 4-dimensional space groups and is well described in the paper of Bülow, Neubüser and Wondratschek [10] and in the subsequent book of Brown, Bülow, Neubüser, Wondratschek and Zassenhaus [8]. The latter book is dedicated to the memory of Hermann whose papers [23], [24] contain the essence of the material described in these notes, but clearly represent only a small fraction of what Hermann knew. The extensions towards n = 5,6,7 have been begun by Ryskov [42] [43] and by Plesken and Pohst [38].

The study of particular classes of space groups in arbitrary dimensions is at present fragmentary. Extensive results using Lie algebra techniques have been obtained by Broderick and Maxwell [7] (see also [33] and [48]). The study of colour symmetry has been bedevilled by a lack of precise definitions when the number of colours is greater than two. The definition used in these notes is that of Burckhardt and van der Waerden [57]; it is unfortunate that this paper was apparently ignored by Shubnikov [51] and Loeb [31] whose books give incomplete and unsystematic listings which are corrected in [28]. Colour symmetry groups not mentioned in these books (for example the colour lattices mentioned at the end of Chapter Five) are quite familiar to metallurgists and physicists in connection with twinning (see Chapter Seven of these notes and Chapter Ten of [30]).

For further historical remarks see [13] and [8]. The above note is based on a quite independent review of the literature but inevitably there is some overlap between accounts from different points of view.

References

[1] Barlow, W., Über die geometrische Eigenschaften homogener starrer Strukturen. Z. Kryst. Min. 23(1894), 1-63.

[2] Barlow, W., Nachtrag zu den tabellen. Z. Kryst. Min. 25 (1896), 86-91.

[3] Bieberbach, L., Über die Bewegungsgruppen der Euklidischen Raüme mit einem endlichen Fundamentalbereich. Gött. Nachr. (1910), 75-84.

[4] Bieberbach, L., Über die Bewegungsgruppen der Euklidischen Raüme. Math. Ann. 70(1910), 297-336 and 72(1912), 400-412.

[5] Brauer, R. and C. Nesbitt, On the modular characters of groups. Ann. Math. 42(1941), 556-590.

[6] Bravais, A, Les systèmes formés par des points distribués regulièrement sur un plan ou dans l'espace. J. Ecole Polytech. 19(1850), 1-128.

[7] Broderick, N. and G. Maxwell, The crystallography of Coxeter groups II. J. Algebra 44(1977), 290-318.

[8] Brown, H., R. Bülow, J. Neubüser, H. Wondratschek and H. Zassenhaus, Crystallographic groups of 4-dimensional space. John Wiley. New York 1978.

[9] Buerger, M. J., Elementary crystallography. John Wiley. New York 1956 and 1963.

[10] Bülow, R., J. Neubüser and H. Wondratschek, On crystallography in higher dimensions. Acta Cryst. A27(1971), 517-535.

[11] Burckhardt, J. J., Bermerkungen zur arithmetischen Berechnung der Bewegungsgruppen. Comm. Math. Helvet. 2(1930), 91-98. Zur theorie der Bewegungsgruppen. Comm. Math. Helvet. 6(1933/34), 159-184. Bewegungsgruppen in mehrdimensionalen Räumen. Comm. Math. Helvet. 9(1936/37), 284-302.

[12] Burckhardt, J. J., Die Bewegungsgruppen der Kristallographie Birkhäuser Verlag. Basel 1946 and 1966.

[13] Burckhardt, J. J., Zur Geschichte der Entdeckung der 230 Raumgruppen. Arch. Hist. Exact. Sci. 4(1967/68), 235-246.

[14] Fedorov, E. S., The elements of the study of figures. Proc. S. Peterb. Mineral. Soc. 21 (1885), 1-289. Symmetry of finite figures. Proc S. Peterb. Mineral. Soc. 25 (1889), 1-52 (Russian)

[15] Fedorov, E. S., Symmetry of regular systems of figures. Proc. S. Peterb. Mineral. Soc. 28 (1891), 1-146. (Russian)

[16] Fedorov, E. S., Symmetry in the plane. Proc. S. Peterb. Mineral. Soc. 28 (1891), 345-389. (Russian)

[17] Fedorov, E. S., Zusammenstellungen der krystallographischen Resultate des Herrn Schönflies und der meiningen. Z. Kryst. Min. 20 (1892), 25-75.

[18] Fedorov, E. S., Theorie der Krystallstruktur. Einleitung. Regelmassiger Punktsysteme. Z. Kryst. Min. 24 (1895), 210-252.

[19] Fedorov, E. S., Theorie der Krystallstruktur. I. Mögliche Structuren. Z. Kryst. Min. 25 (1896), 113-224

[20] Fedorov, E. S., Symmetry and structure of crystals. Akad. Nauk SSSR. Moscow 1949 (Russian). Translation published by American Crystallographic Association 1971.

[21] Fricke, R. and F. Klein, Vorlesungen uber die Theorie der Automorphen Functionen I. Teubner. Leipzig 1897.

[22] Gray, J., Olinde Rodrigues' 1840 paper on transformation groups Arch. Hist. Exact Sci. (to appear)

[23] Hermann, C., Kristallographie in Raümen beliebiger Dimensionzahl. I Die Symmetrieoperationen. Acta Cryst. 2 (1949), 139-145.

[24] Hermann, C., Translationsgruppen in n-Dimensionen. Zur Struktur und Materie der Festkörper. Springer. Berlin 1952.

[25] Hessel, J. F. C., Krystall. Gehler's Physikalische Wörterbuch Schwickert, Leipzig 1830 reprinted in Ostwald's Klassiker der Exakten Wissenschaften Engelmann. Leipzig 1897.

[26] Hilbert, D., Mathematical problems. Gött. Nachr (1900) 253-297 translated in Bull. Amer. Math. Soc. 8 (1902) 437-479.

[27] Jarratt, J. D., The Decomposition of Crystal Families (to appear)

[28] Jarratt, J. D. and R. L. E. Schwarzenberger, Coloured plane groups (to appear)

[29] Jordan, C., Sur les groupes de mouvements. Ann. mat. pur app 2 (1869), 167-215 and 322-345. Collected Works IV pp 231-301.

[30] Kelly, A. and G. W. Groves, Crystallography and crystal defects. Longmans. London 1970.

[31] Loeb, A. L., Color and symmetry. John Wiley. New York 1971.

[32] Macdonald, S. O. and A. P. Street, The seven friezes and how to colour them. Utilitas Math. 13 (1978), 271-292.

[33] Maxwell, G., The crystallography of Coxeter groups. J. Algebra 35 (1975) 159-177.

[34] Minkowski, H., Zur Theorie der positiven quadratischen Formen. Crelle J. reine v. angew. Math. 106 (1890) 5-26. Collected Works I, pp. 212-218.

[35] Möbius, A. F., Uber das Gesetz der Symmetrie der Kristalle. Crelle J. reine v. angew. Math. 43 (1852) 365 and Uber symmetrische Figuren. Crelle J. rein v. angew Math, 44 (1852) 335. Collected Works II, pp. 349-372.

[36] Neubüser, J., W. Plesken and H. Wondratschek. An emendatory discussion on defining crystal systems.

[37] Niggli, P., Die Flächensymmetrien homogenes Diskontinuen. Z. Krist. 60 (1924), 283-298.

[38] Plesken, W. and M. Pohst, On maximal finite irreducible subgroups of GL(n,Z). Math. Computation 31 (1977), 536-573.

[39] Polya, G., Uber die Analoge der Kristallsymmetric in der Ebene. Z. Krist. 60 (1924), 278-282.

[40] Polya, G., Kombinatorische Anzahlbestimmingen für Gruppen, Graphen und chemische Verbindungen. Acta Math. 68 (1937), 145-254.

[41] Rodrigues, O., Des lois géométriques qui régissent les déplacements d'un ststeme solide dan l'espace, et de la variation des coordonées provenant de ces deplacements considérés indépendamment des causes qui penvent les produire. Jour. de Math. 5 (1840), 380-440.

[42] Ryskov, S. S., On maximal finite groups of integer n × n matrices. Dokl. Akad Nank SSSR. 204 (1972), 561-564 translated in Sov. Math. Dokl. 13 (1972), 720-724.

[43] Ryskov, S. S., Maximal finite groups of integral n × n-matrices and full groups of automorphisms of positive quadratic forms (Bravais models). Trudy. Mat. Inst. Steklov 128 (1972), 183-211 translated in Proc. Steklov. Inst. Math. 128 (1972), 217-250.

[44] Schönflies, A. M., Uber Gruppen von Bewegungen. Math. Ann. 28 (1886), 319-342 and 29 (1887), 50-80.

[45] Schönflies, A. M., Uber Gruppen von Transformationen des Raumes in sich. Math. Ann. 34 (1889), 172-203.

[46] Schönflies, A. M., Kristallsysteme und Kristallstruktur. Teubner. Leipzig 1891. Second edition Theorie der Kristallstruktur. Berlin 1923.

[47] Schwarzenberger, R. L. E., Classification of crystal lattices. Proc. Camb. Philos. Soc. 72 (1972), 325-349.

[48] Schwarzenberger, R. L. E., Crystallography in spaces of arbitrary dimension. Proc. Camb. Philos. Soc. 76 (1974), 23-32.

[49] Schwarzenberger, R. L. E., The 17 plane symmetry groups. Math. Gazette 58 (1974), 123-131.

[50] Schwarzenberger, R. L. E., Graphical representation of n-dimensional space groups. (to appear)

[51] Shubnikov, A. V. and N. V. Belov and others. Coloured symmetry. Pergammon Press. Oxford 1964.

[52] Shubnikov, A. V. and V. A. Kopstik, Symmetry in Science and Art. Plenum New York 1974.

[53] Sohncke, L., Entwicklung einer Theorie der Kristallstruktur Teubner. Leipzig 1879.

[54] Sohncke, L., Erweiterung der Theorie der Krystallstruktur Z. Kryst. Min 14 (1888), 426-446.

[55] Sohncke, L., Bermerkungen zu Herrn Wulff's Theorie der Krystallstruktur. Z. Kryst. Min. 14 (1888), 417-425. Die Struktur der optisch drehenden Krystalle. Z. Kryst. Min. 19 (1891) 529-559.

[56] van der Waerden, B. L., Die Reduktionstheorie der positiven quadratischen Formen. Acta. Math. 96 (1956), 265-293.

[57] van der Waerden, B. L. and J. J. Burckhardt, Farbgruppen. Z. Krist. 115 (1961), 231-234.

[58] Weyl, H., Symmetry. Princeton University Press 1952.

[59] Wolf, J. A., Spaces of constant curvature. McGraw-Hill. New York 1967.

[60] Zassenhaus, H., Neuer Beweis der Endlichkeit der Klassenzahl bei unimodularer Aquivalenz endlicher ganzzahliger substitutionsgruppen. Abh. Math. Sem. Hamburg 12 (1938), 276-288.

[61] Zassenhaus, H., Uber einen Algorithmus zur Bestimmung der Raumgruppen. Comm. Math. Helvet. 21 (1948), 117-141.